Hydraulic Rubber Dam

PLASTICS DESIGN LIBRARY (PDL) PDL HANDBOOK SERIES

Series Editor: Sina Ebnesajjad, PhD (sina@FluoroConsultants.com) President, FluoroConsultants Group, LLC Chadds Ford, PA, USA
http://www.FluoroConsultants.com

The **PDL Handbook Series** is aimed at a wide range of engineers and other professionals working in the plastics industry, and related sectors using plastics and adhesives.

PDL is a series of data books, reference works and practical guides covering plastics engineering, applications, processing, and manufacturing, and applied aspects of polymer science, elastomers and adhesives.

Recent titles in the series

Introduction to Plastics Engineering, Shrivastava (ISBN: 9780323395007)
The Effect of Sterilization Methods on Plastics and Elastomers, McKeen (ISBN: 9780128145111)
Polymeric Foams Structure-Property-Performance, Obi (ISBN: 9781455777556)
Technology and Applications of Polymers Derived from Biomass, Ashter (ISBN: 9780323511155)
Fluoropolymer Applications in the Chemical Processing Industries, 2e, Ebnesajjad & Khaladkar (ISBN: 9780323447164)
Reactive Polymers, 3e, Fink (ISBN: 9780128145098)
Service Life Prediction of Polymers and Plastics Exposed to Outdoor Weathering, White, White & Pickett, (ISBN: 9780323497763)
Polylactide Foams, Nofar & Park (ISBN: 9780128139912)
Designing Successful Products With Plastics, Maclean-Blevins (ISBN: 9780323445016)
Waste Management of Marine Plastics Debris, Niaounakis (ISBN: 9780323443548)
Film Properties of Plastics and Elastomers, 4e, McKeen (ISBN: 9780128132920)
Anticorrosive Rubber Lining, Chandrasekaran (ISBN: 9780323443715)
Shape-Memory Polymer Device Design Safranski & Griffis, (ISBN: 9780323377973)
A Guide to the Manufacture, Performance, and Potential of Plastics in Agriculture, Orzolek (ISBN: 9780081021705)
Plastics in Medical Devices for Cardiovascular Applications, Padsalgikar (ISBN: 9780323358859)
Industrial Applications of Renewable Plastics, Biron (ISBN: 9780323480659)
Permeability Properties of Plastics and Elastomers, 4e, McKeen (ISBN: 9780323508599)
Expanded PTFE Applications Handbook, Ebnesajjad (ISBN: 9781437778557)
Applied Plastics Engineering Handbook, 2e, Kutz (ISBN: 9780323390408)
Modification of Polymer Properties, Jasso-Gastinel & Kenny (ISBN: 9780323443531)
The Science and Technology of Flexible Packaging, Morris (ISBN: 9780323242738)
Stretch Blow Molding, 3e, Brandau (ISBN: 9780323461771)
Chemical Resistance of Engineering Thermoplastics, Baur, Ruhrberg & Woishnis (ISBN: 9780323473576)
Chemical Resistance of Commodity Thermoplastics, Baur, Ruhrberg & Woishnis (ISBN: 9780323473583)
Color Trends and Selection for Product Design, Becker (ISBN: 9780323393959)
Fluoroelastomers Handbook, 2e, Drobny (ISBN: 9780323394802)
Introduction to Bioplastics Engineering, Ashter (ISBN: 9780323393966)
Multilayer Flexible Packaging, 2e, Wagner, Jr. (ISBN: 9780323371001)
Fatigue and Tribological Properties of Plastics and Elastomers, 3e, McKeen (ISBN: 9780323442015)
Emerging Trends in Medical Plastic Engineering and Manufacturing, Schönberger & Hoffstetter (ISBN: 9780323370233)
Manufacturing and Novel Applications of Multilayer Polymer Films, Langhe & Ponting (ISBN: 9780323371254)
Recycling of Polyethylene Terephthalate Bottles, Thomas (ISBN: 9780128113615)

To submit a new book proposal for the series, or place an order, please contact Edward Payne, Acquisitions Editor at edward.payne@elsevier.com

Hydraulic Rubber Dam

An Effective Water Management Technology

Edited by

Sabu Thomas

School of Chemical Sciences, Mahatma Gandhi University, Kottayam, India
International and Interuniversity Centre for Nanoscience and Nanotechnology, Mahatma Gandhi University, Kottayam, India

Ajay Vasudeo Rane

Composite Research Group, Department of Mechanical Engineering, Durban University of Technology, Durban, South Africa

Abitha V.K.

School of Chemical Sciences, Mahatma Gandhi University, Kottayam, India

Krishnan Kanny

Composite Research Group, Department of Mechanical Engineering, Durban University of Technology, Durban, South Africa

Aastha Dutta

Plastics and Polymer Engineering Department, G.S.Mandal's Maharashtra Institute of Technology, Aurangabad, Maharashtra, India

ELSEVIER

William Andrew
Applied Science Publishers

William Andrew is an imprint of Elsevier
The Boulevard, Langford Lane, Kidlington, Oxford, OX5 1GB, United Kingdom
50 Hampshire Street, 5th Floor, Cambridge, MA 02139, United States

British Library Cataloguing-in-Publication Data
A catalogue record for this book is available from the British Library

Library of Congress Cataloging-in-Publication Data
A catalog record for this book is available from the Library of Congress

ISBN: 978-0-12-812210-5

For Information on all William Andrew publications
visit our website at https://www.elsevier.com/books-and-journals

Publisher: Matthew Deans
Acquisition Editor: Edward Payne
Editorial Project Manager: Mariana L. Kuhl
Production Project Manager: Kamesh Ramajogi
Cover Designer: Miles Hitchen

Typeset by MPS Limited, Chennai, India

Contents

List of Contributors ix

1 **Rubber Dam—An Introduction** **1**
Ajay Vasudeo Rane, Surendra Thakur and Rookmoney Thakur

References 9

2 **Water Conservation Technologies** **11**
Arun Kumar

2.1 Introduction 11
2.2 Conservation of Surface Water Resources 12
2.3 Conservation of Groundwater Resources 12
2.4 Rainwater Harvesting 13
2.5 Rubber Dam for Water Conservation and Management 13
2.6 River Morphology 14
2.7 Rubber Dam Storage 15
2.7.1 Capacity and Filling Time 15
2.7.2 Seepage Losses From the Pond 15
2.7.3 Evaporation Losses From the Pond 16
2.8 Typical Installations of Rubber Dam for Water Conservation 16
2.8.1 Linyi City of Shandong State in China 16
2.8.2 Nakdonggang River in the Daegu City of South Korea 16
2.8.3 Musi River in the City of Hyderabad, India 19
2.8.4 Artificial Recharge of Groundwater From Rubber Dams 19
2.9 Fluid Adopted for Rubber Dam 20
2.10 Aerators for Improving Water Quality 20
2.11 Ecological Manifestation 22
2.11.1 Producers (Phytoplankton) 22
2.11.2 Consumers (Zooplankton, Etc.) 22
2.11.3 Various Animals and Other 23

2.12 Impact on Environment and Biodiversity 23
2.13 Impact on Society 23
References 23

3 Raw Materials—Rubber Dam 25
Abitha V.K., Sabu Thomas, Ajay Vasudeo Rane and Krishnan Kanny

3.1 Introduction 25
References 36

4 Manufacturing Process—Reinforced Rubber Sheet for Rubber Dam 37
Ajay Vasudeo Rane and Krishnan Kanny

4.1 Introduction 37
References 46

5 Bonding and Anchoring System for Rubber Dam 47
Joygopal Jena

5.1 Introduction 48
5.2 Category of Rubber Dams 49
5.2.1 According to Medium of Fill 49
5.2.2 According to the Side Slope 49
5.2.3 According to the Method of Anchoring 51
5.2.4 Classification Based on the Seam of Rubber Bag 52
5.3 Bonding of Dam Body 53
5.3.1 Specification of Rubber Dam 54
5.3.2 Physical Property of Rubber Dam 54
5.3.3 Considerations for Construction of a Rubber Dam Body 55
5.3.4 Durability of the Rubber Dam 56
5.4 Anchoring System 56
5.4.1 Anchorage of Dam Bag With Concrete Base 56
5.4.2 Anchorage of Rubber Dam Sheets With Side Walls of Abutment or Piers 57
5.5 Various Types of Anchoring 58
5.5.1 Concrete Wedge Blocks 58
5.5.2 Steel Clamps With Anchor Bolts 58

5.6 Clamping Rubber Fabric With Concrete 59
5.6.1 Clamping by Punching the Fabric 62
5.6.2 Clamping Without Punching the Fabric 62
5.6.3 Clamping Plate and Anchor Bolts 62
5.7 Foundation for the Rubber Dam 62
5.7.1 Loads Considered for the Design of the Concrete Base 63
5.7.2 Selection of Site for Suitability of Foundation 63
5.7.3 General Construction Procedure for Lying of Foundation Concrete 63
5.8 Special Types of Rubber Dams 64
5.8.1 Shield-Type Rubber Dam 64
Acknowledgments 65
References 65
Further Reading 66

6 Inflation and Deflation of Rubber Dam 67
Arun Kumar and Shahid ul Islam

6.1 Introduction 67
6.2 Parts of Rubber Dam 69
6.3 Medium of Inflation 70
6.4 Advantages of Rubber Dam 71
6.5 Installation Method 72
6.6 Inflation and Deflation Phenomenon of Rubber Dam 74
6.7 Overflow Above and Deflated Flexible Membrane Dam 77
6.7.1 Introduction 77
6.7.2 Overflow of Inflated Rubber Dams 78
6.7.3 Coefficient of Discharge 78
6.7.4 Maintenance During Inflation and Deflation 81
6.8 Design Criteria for Rubber Dam System 83
6.8.1 Inflation 84
6.8.2 Deflation 84
6.8.3 Emergency Mechanical Deflation System 85
6.8.4 Design Flow Event With Rubber Dam Bladder Deflated 85
6.8.5 Rubber Dam Bladder Specification 86
6.8.6 Rubber Dam Bladder Anchorage 86

6.8.7 Operational Equipment 87
6.8.8 Rubber Dam Bladder, Anchoring, and Pipe Connections 89
6.9 Seismic National and International Requirements 89
6.10 Operational Steps for Water-Filled Dam 89
6.11 Inflation Time 90
6.12 Optimization of Suitable Inflation Deflation Mechanism 91
6.13 Safe and User-Friendly Operation 91
6.14 Reference Standards 93
6.15 Conclusion 95
References 96

7 Life Cycle Assessment of Rubber Dam 99
Shahid ul Islam and Arun Kumar

7.1 Introduction 99
7.2 Life Cycle Assessment 102
7.3 Objectives of Life Cycle Assessment 103
7.4 Life Cycle Inventory Analysis 103
7.5 Life Cycle Cost 104
7.6 Case Study 105
7.7 Life Cycle Assessment Comparison with Conventional Weirs 107
7.7.1 Results 109
7.7.2 Reference Standards 111
7.8 Conclusions 120
References 122

Index 123

List of Contributors

Abitha V.K. School of Chemical Sciences, Mahatma Gandhi University, Kottayam, India

Shahid ul Islam Baba Ghulam Shah Badshah University, Rajouri, India

Joygopal Jena Department of Civil Engineering, Gandhi Institute of Technological Advancement (GITA), Bhubaneswar, India

Krishnan Kanny Composite Research Group, Department of Mechanical Engineering, Durban University of Technology, Durban, South Africa

Arun Kumar Alternate Hydro Energy Centre, Indian Institute of Technology, Roorkee, India

Ajay Vasudeo Rane Composite Research Group, Department of Mechanical Engineering, Durban University of Technology, Durban, South Africa

Rookmoney Thakur International Centre of Non-Violence, Durban University of Technology, Durban, South Africa

Surendra Thakur Centre for Continuous & Professional Education, Durban University of Technology, Durban, South Africa

Sabu Thomas School of Chemical Sciences, Mahatma Gandhi University, Kottayam, India; International and Interuniversity Centre for Nanoscience and Nanotechnology, Mahatma Gandhi University, Kottayam, India

1 Rubber Dam—An Introduction

Ajay Vasudeo Rane[1], Surendra Thakur[2] and Rookmoney Thakur[3]
[1]Composite Research Group, Department of Mechanical Engineering, Durban University of Technology, Durban, South Africa, [2]Centre for Continuous & Professional Education, Durban University of Technology, Durban, South Africa, [3]International Centre of Non-Violence, Durban University of Technology, Durban, South Africa

Abstract

Former United Nations Secretary-General Boutros Ghali (1985) warned: "the wars in the future are likely to be fought over water." For centuries, warfare and conflict has been interrelated to the protection of water resources. Water is a source of life. And like oil and other natural resources, it follows that the availability of water is every bit as vital to human improvement as energy security, however, with one huge limitation: unlike oil, water has no known substitutes. As such, no nation can afford to suffer a calamitous loss of water resources. For years experts have warned that this limited resource is fast becoming more critical with the changing requirement of an expanding global populace and effects of climate change around its availability. Many countries in the Middle East, Africa, Central and South Asia—for example, Afghanistan, Pakistan, China, Kenya, Egypt, and India—are already feeling the direct consequences of the water scarcity—with the competition for water leading to social unrest, conflict, and migration. Water is an essential entity to sustain human life and other living organism. The conflict between human population and earth's unchanging supply of fresh water is worsening every year, which possess a challenge in meeting the water needs of an increasing human population while parting an amount for environmental life cycles. To overcome the problem of water shortage "dams" are an appropriate solution for water conservation.

Keywords: Rubber dam; air filled; water filled; check dam

Former United Nations Secretary-General Boutros Ghali (1985) warned: "the wars in the future are likely to be fought over water." For centuries warfare and conflict has been interrelated to the protection of water

Hydraulic Rubber Dam. DOI: https://doi.org/10.1016/B978-0-12-812210-5.00001-8

resources (Shiva, 2016). Water is a source of life. And like oil and other natural resources, it follows that the availability of water is every bit as vital to human improvement as energy security, however, with one huge limitation: unlike oil, water has no known substitutes. As such, no nation can afford to suffer a calamitous loss of water resources. For years experts have warned that this limited resource is fast becoming more critical with the changing requirement of an expanding global populace and effects of climate change around its availability. Many countries in the Middle East, Africa, Central and South Asia—for example, Afghanistan, Pakistan, China, Kenya, Egypt, and India—are already feeling the direct consequences of the water scarcity—with the competition for water leading to social unrest, conflict and migration.

> *The one idea that standout for its simplicity, efficacy and affordability is rain water harvesting. Capture rain water, store it and use it—it is as simple that. If appropriate technologies are built around this simple idea, they can provide decentralized, local—level solutions that can considerably meet the drinking water needs of our urban and rural populations.*
>
> *– Atal Bihari Vajpayee, Prime Minister of India, 2000 as cited in the book of Patrick McCully, Silenced Rivers (McCully, 2001).*

As mentioned earlier, water is one of the most important substances on earth. All plants and animals must have water to survive; if there was no water, there would have been no life on earth (http://www.health.gov.au/internet/publications/publishing.nsf/Content/ohpenhealth-manual-atsi-cnt-l~ohp-enhealth-manual-atsi-cnt-l-ch6~ohp-enhealth-manual-atsi-cnt-l-ch6.1). Water is an essential entity to sustain human life and other living organisms. The conflict between human population and the earth's unchanging supply of fresh water is worsening every year which possess a challenge in meeting the water needs of an increasing human population while parting an amount for environmental life cycles. To overcome the problem of water shortage, "dams" are an appropriate solution for water conservation. The history of dams has been traced to about 2900 BC when a 49-ft high masonry structure was built on the Nile at Kosheish to supply water to King Mene's capital at Memphis. The oldest dam still in use is a rock fill structure about 20 ft high on Orontes in Syria, built during 1300 BC (http://en.banglapedia.org/index.php?title = Dam). The word dam has synonyms like barrage, barrier, wall, embankment, levee, barricade, obstruction, hindrance, and blockage. In water conservation—a dam is a barrier that stops and restricts the flow of water or underground streams, that is, to collect water which is then evenly distributed between locations. The primary purpose of dams is to retain water while structures like levees prevent water flow into specific land regions. Dams are typically categorized as per Fig. 1.1. Fig. 1.2 provides the detailed classification of dams, based on structure, use, and materials. Fig. 1.3 classifies the function of dams.

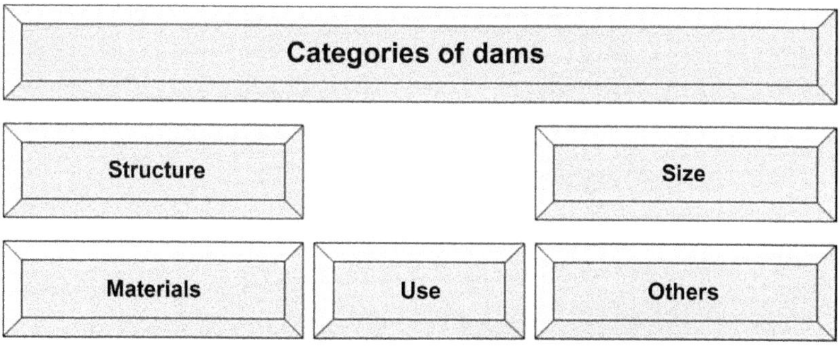

Figure 1.1 Categories of dams (https://en.wikipedia.org/wiki/Dam).

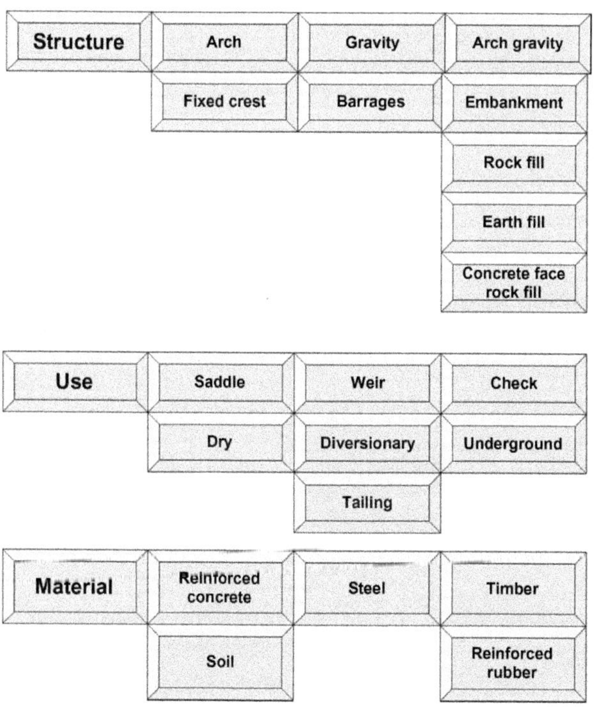

Figure 1.2 Classification of dams on basis of their structure, use, and material (https://en.wikipedia.org/wiki/Dam).

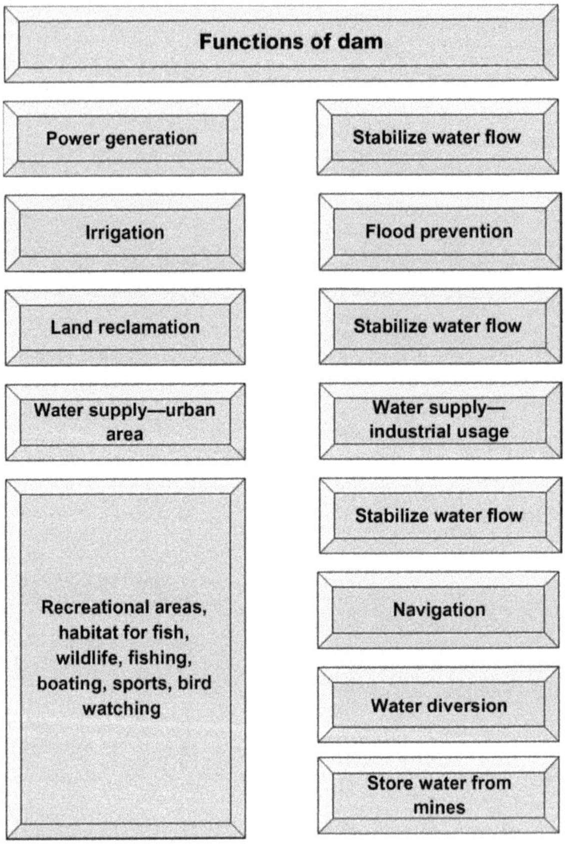

Figure 1.3 General functions of dams (https://en.wikipedia.org/wiki/Dam).

With the development of polymer composites, there is a new type of hydraulic structures, which is the inflatable rubber dam belongs to the flexible wall structure. Inflatable rubber dams have been used as water control structures for more than 50 years. The world's first inflatable rubber dam was installed in Los Angeles Country in the United States in the mid-1950s. In those days, inflatable rubber dams were viewed much as they are today as relatively inexpensive versatile structures capable of creating pondage and/or controlling flow in a particular water course (Technical Manual). Rubber dam is long tubular-shaped fabric (coated with rubber) and sand witched between rubber sheets and used as a water control structure since 1950. Rubber dams are seamless structures fixed to a reinforced concrete using accessories like clamp plates and anchor bolts, the rubber dams are then inflated with air or water as per designed (for pressure and height) to raise the upstream water level (https://en.wikipedia.org/wiki/Dam). Rubber dams has numerous advantages over other types of water control gate, such as simple and inexpensive operating system, can be installed on almost any channel cross-section shape, relatively low capital cost, perfect sealing, that is, no leakage, virtually maintenance free, that is, no moving parts and no painting, long spans up to 100 m and multiple spans of several hundred meters, easily designed to accept loading in both directions, light structure, clean operation (no hydraulic required), can always be opened (deflated) and no possibility of jamming (Technical Manual).

The construction cost of the inflatable rubber dam is low. According to information available, the average investment can be reduced by 30%–70%. The construction period is not very long, and most of the inflatable rubber dams can be completed in the year of construction (http://www.rubberdam.org/product/water-filled-rubber-dam.html). As the inflatable rubber dam has a flexible structure and is economical; hence, it can effectively buffer the impact of water, resist earthquakes, and so on (http://www.rubberdam.org/product/water-filled-rubber-dam.html). It can also be applied to garden projects in urban areas and has a wide range of applications to retain water, divert water for irrigation, create pondage, control flood, control water, erosion control, tidal barriers, raise existing dams, develop water entertainment, power generation, urban landscaping, ground water recharging, sewage treatment, adjust water level, generate hydroelectric power, shipping, increase reservoir capacity, ecological protection, antitide, erosion control, river regulation prevention of overtopping, and chemical spillage in streams (Technical Manual).

The key element of the inflatable rubber dam is the rubber body (Technical Manual). Rubber dams are made of multilayer fabric comprising strength, canvas, and rubber layers, which consist of outer, middle, and inner layers which serves their function during field performance. Rubber compound for outer layer is formulated to impart excellent ultraviolet (UV), heat, ozone, and abrasion resistance. The outer rubber layer has excellent weathering, aging, abrasion, sunlight, UV, heat, and ozone resistance. The middle rubber layer is used for protection of canvas and for connection of two or more canvas with rubber layers. The inner layer is used for protection of canvas and has excellent watertight and airtight properties. The canvas has great flexural stress and adhesion function (http://www.rubberdam.org/product/water-filled-rubber-dam.html).

Commercial classification for rubber dam on the medium of inflation and deflation, and structure is done in Fig. 1.4. Air-filled rubber dam, as the name suggests, indicates a rubber bladder, in which air is pumped

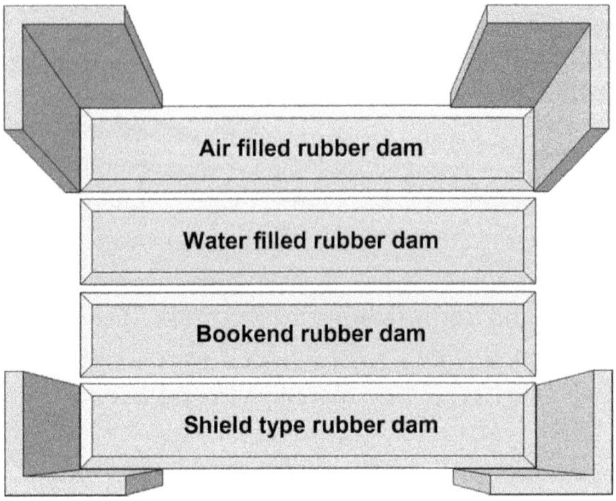

Figure 1.4 Commercial classification of inflatable rubber dam (http://www.rubberdam.org/product/water-filled-rubber-dam.html).

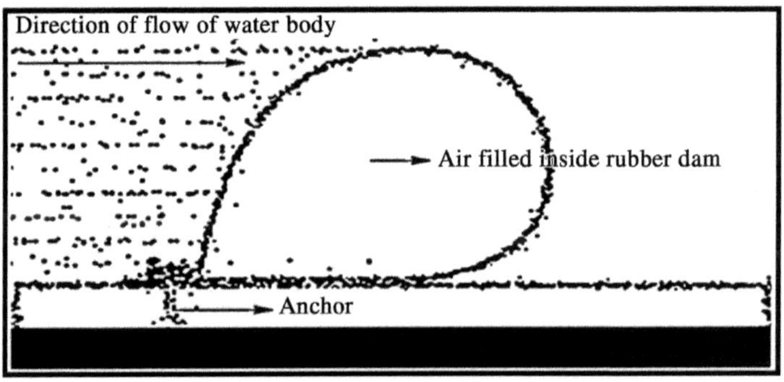

Figure 1.5 Cross-section of air-filled rubber dam.

to inflate (see Fig. 1.5). Air-filled rubber dams are used across channels; streams to store water or divert water for irrigation purpose. Air-filled rubber dams are economic (low cost), as they do not make use of high-cost water pumps, piping, and wide concrete foundation for specified applications. Inflation and deflation are quicker, hence consume less power for their operation. As air is pumped in and out to inflate and deflate, it eliminates the risk of freezing in cold weather condition (http://www.rubberdam.org/product/water-filled-rubber-dam.html). Water-filled rubber dam, are filled with stabilizing water (see Fig. 1.6). In comparison to air-filled rubber dams—high operation costs are involved for water-filled rubber dam as longer times are required for inflation and deflation.

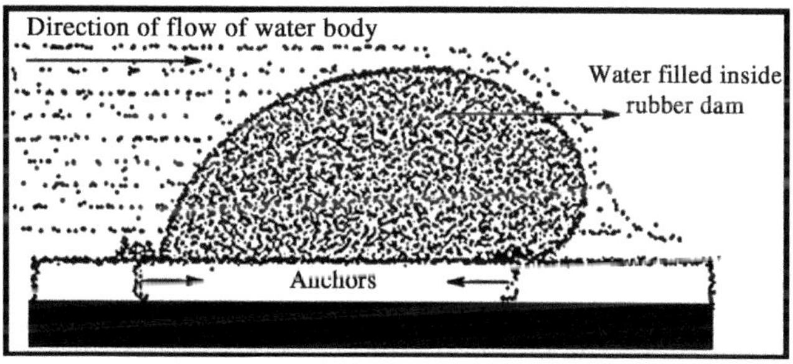

Figure 1.6 Cross-section of water filled rubber dam.

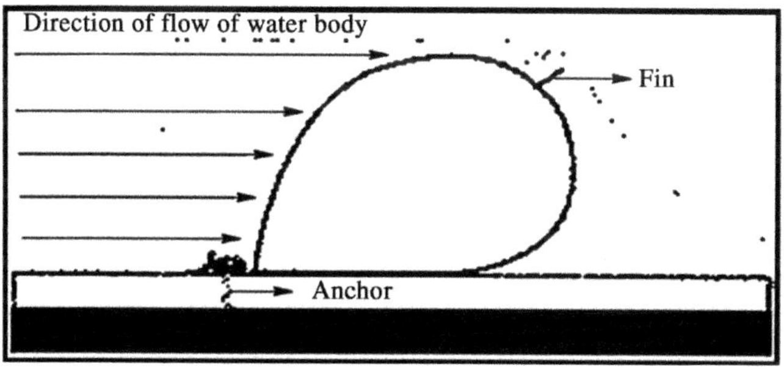

Figure 1.7 Cross-section of bookend rubber dam.

Maintenance cost for water-filled rubber dam is low. Water-filled rubber dams are steady providing control over upstream water levels in comparison to air-filled rubber dams (http://www.rubberdam.org/product/water-filled-rubber-dam.html). There is a high risk for freezing in winter; hence, water circulation is needed. Bookend rubber dam is finned (see Fig. 1.7) which forms a rising angle after inflation with air or water. Fin reduces the vibration and abrasion of the dam by eliminating vacuum underneath the bookend rubber dam. Bookend rubber also possesses excellent lateral tension. It has resistance against dirt and sediment. Shield-type rubber dam integrates structure of traditional steel dam and rubber dam. Structures composed of air-filled rubber bladder, shielded with steel structures (front plates and support plates), are seen in Fig. 1.8. Shield-type rubber dams are easy to install, easy to repair, longer service life, multilevel application (http://www.rubberdam.org/product/water-filled-rubber-dam.html). Design consideration plays an important role in serviceability of rubber dam installed for intended application, for example: outer layer of rubber dam should be resistant to damage by impact of sharpen bits and pieces, if not a cover layer should be a part of assembly of component during its

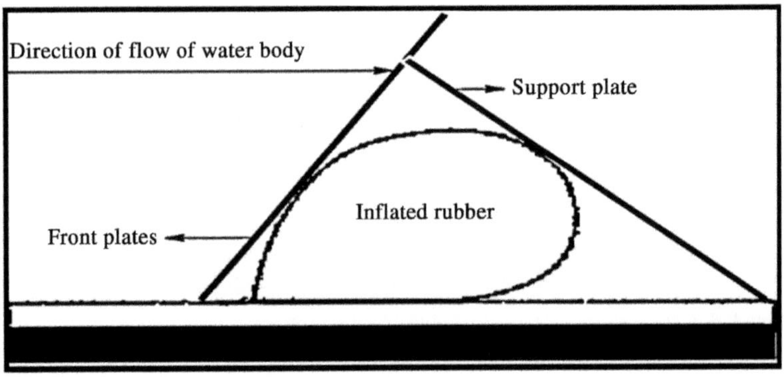

Figure 1.8 Side view of shield-type rubber dam.

initial manufacturing stages. Considering the clamping system—double clamp systems provide strength to rubber structure than single clamping system. Angle of inclination is an important parameter while setting up a rubber dam on a field, which determines the wrinkle formation during inflation and deflation, wrinkle formation during operation should be minimized to increase the life span of rubber dam. Filling medium, for example, water-filled rubber dams are more stable than air-filled ones. In deflated condition, rubber dam must lie flat on surface, minimizing disturbance to flow. Fitment clamps and anchors must be aligned and fitted skillfully to avoid leakage of filled media (air or water) during its operation (Pardeshi and Mahajan). Natural and synthetic rubbers compounded with specific additives (fillers, accelerators, process aids), coated fabric is used as a reinforcement in between the rubber sheets, enhances the strength of rubber dam during application. More details on raw materials and process involved in fabrication of rubber dam are discussed further in the book. Clamping plates and anchor bolts are used in fitment during construction of a rubber dam structure in desired location. Smooth functioning of rubber dam is required, for which a control system involving pressure gauge, water gauge, pressure regulator, inflate and deflate valve, check valve and float valve is made use of, now a days automatic control systems are also available. Information on bonding, anchoring, inflation and deflation in rubber dams, effect of different stress on rubber dam, and life cycle assessment of rubber dam is also part of this book. To conclude, rubber dams, as water harvesting structures, will help in management of natural resources and conservation of water for agrarian countries that needs higher number of small dams. Construction of rubber dams in place of traditional dam enhances the irrigation efficiency and cost effectiveness. Rubber dams, as check dams, are capable to prevent flood, reduce dependency on monsoon, increase in level of water storage for release, and uniform distribution during droughts leads to efficient use and conservation of water.

References

McCully, P., 2001. Silenced Rivers – The Ecology and Politics of Large Dams. ZED Books, London and New York, ISBN – 1856499014.

Pardeshi, D.G., Mahajan, K.S., 2016. Rubber dam—effective method to improve upstream flow of river. J. Water Resour. Pollut. Stud. 1 (2), 1–14.

Shiva, V., 2016. Water Wars: Privatization, Pollution, and Profit. North Atlantic Books, Berkeley, CA.

Technical Manual. Spillway Gates and Inflatable Rubber Dams. Dyrhoff Ltd.

2 Water Conservation Technologies

Arun Kumar
Alternate Hydro Energy Centre, Indian Institute of Technology, Roorkee, India

Abstract

Rubber dams are used for water conservation in different parts of the world due to their technical advantage and operation ease. The stored water behind the rubber dam is used for recharging the groundwater, storing the freshwater, improving the landscape, and enhancing the fisheries.

Keywords: Freshwater; groundwater; water conservation; aquatic ecosystem; water resources; aerator

2.1 Introduction

Rapid industrialization and urbanization coupled with continuous decline in per capita water availability is putting a lot of pressure on the available water resources in the world. The increasing gap between water availability and demand highlights the need for conservation of water. The different Government Water Policies also lay stress on conservation of water. It has been stipulated that efficiency of utilization in all the diverse uses of water should be optimized and an awareness of water as a scarce resources should be fostered.

Due to water withdrawal at different locations on rivers for irrigation and drinking purposes, the rivers become almost dry specially during the lean months. Some of the rivers carry treated or untreated industrial as well as municipal wastewater, making river bad to worst. The general hydrology of many rivers shows that rivers have low discharge to high discharge during the year.

There is a need for water conservation, not only to restore the fast deteriorating ecosystem of the rivers in different parts of the world but also to meet thc inevitable emergency of shortage even for drinking and irrigation in near future. To plan strategies for meeting the water shortage, following may be considered CWC (2005):

1. Water resources are theoretically "renewable" through hydrological cycle. However, what is renewable is not only

Hydraulic Rubber Dam. DOI: https://doi.org/10.1016/B978-0-12-812210-5.00002-X

the quantity but also pollution, contamination, climate change, temporal, and seasonal variations have affected the water quality and reduced the amount of "usable water."

2. The groundwater levels have been declining rapidly due to excessive withdrawal.
3. Rainfall is highly unevenly distributed over time and space in various parts of the country.
4. Increased demand in coastal areas is threatening the freshwater aquifers with seawater intrusion.
5. In inland saline areas, the freshwater is becoming saline due to excessive withdrawal of groundwater.
6. Water conservation practices in urban areas can reduce the demand as much as by one-third, in addition to minimizing pollution of surface and groundwater resources.
7. Watershed programs tended to concentrate on harvesting rainwater through surface structures. There is a need to look at surface and groundwater holistically and prepare a conjunctive use plan.

2.2 Conservation of Surface Water Resources

All efforts have to be made to fully utilize the rain runoff and store at all probable storage sites. In addition to creating new storages, it is essential to renovate the existing tanks and water bodies by desilting and repairs. The revival of traditional water-storage techniques and structures should also be given due priority.

2.3 Conservation of Groundwater Resources

Groundwater is an important component of hydrological cycle. It supports the springs in hilly regions and the river flow of all peninsular rivers during the nonmonsoon period. For sustainability of groundwater resources, it is necessary to arrest the groundwater outflows by:

1. construction of subsurface dams,
2. watershed management,
3. treatment of upstream areas for development of springs, and
4. skimming of freshwater outflows in coastal areas and islands.

2.4 Rainwater Harvesting

Rainwater harvesting is the technique of collection and storage of rainwater at the surface or in subsurface aquifers, before it is lost as surface runoff. Groundwater augmentation through diversion of rainfall to subsurface reservoirs, by various artificial recharge techniques, has special relevance in India where due to terrain conditions most of the rainwater is lost as flash floods, and local streams remain dry for most part of the year. Check dams, recharge shafts, dugwell recharge, subsurface dykes, and adopting rooftop rainwater harvesting in urban areas can go a long way in redeeming the worsening situation of shortage of groundwater.

Uncovered areas, particularly in urban and semiurban localities, are continuously diminishing due to phenomenal pace of industrialization and urbanization and massive use of concrete all around in the country. This phenomenon is constantly causing reduced scope for percolation of rainwaters to the ground during monsoon and thus perpetual reduction in groundwater recharge year after year. With a view to offset this loss in recharge of groundwater, there is apparent need for making roof rainwater harvesting mandatory, either through legislation or by promulgating ordinance, for every public as well as private new and existing buildings in urban and semiurban areas within specified time frame. Apart from this, harvesting of surface runoff in open areas, both public and private, may also need to be encouraged.

2.5 Rubber Dam for Water Conservation and Management

Water conservation in the rivers may be planned by way of construction of very low height flexible weirs in the river. The rubber dams/inflated weir/flexible weir may be deflated/inflated during monsoon/lean season regulated/controlled automatically with rugged noncontact ultrasonic water level sensors. The low height flexible weir can be installed at regular interval of river stretch specially designed for the objectives of storage of water (monsoon or from connecting river/canals during nonagricultural period), ensuring minimum flow of water downstream and navigation, recharging groundwater, ensuring ponding for fisheries and power generation, and improving water quality in terms of reduced biological oxygen demand (BOD) and increased dissolved oxygen (DO).

Normally sufficient water is available in the river during rainy period which can be stored easily without affecting any possible scheme. The height of the rubber dam may be planned 2–4 m from the techno economic consideration of the rubber dam and keeping in view the river slope.

The rapid increase in the density of human population in certain pockets of the country as a result of urbanization and industrialization is making adverse impact on the quality of both surface and groundwater. Demand for water is increasing on one hand, and on the other hand, the quantity of "utilizable water resources" is decreasing due to human intervention in the form of pollution of freshwater. Thus, the protection of existing water resources from pollution is a very vital aspect of water conservation.

To cope with water scarcity, time has come to bring the change in the prevailing agriculture practice by using the treated wastewater (of desired or lesser than desired level) and adopting water-efficient irrigation practice like sprinkler and drip, thus reducing the consumptive use of freshwater and making available freshwater to dry river during the lean month for improving esthetic, environment, health condition, and navigation. Such investment on adopting water-efficient irrigation practices along with capacity building, awareness creation, and possible changing cropping pattern may be adopted as per recommendations of agriculture experts.

2.6 River Morphology

Only rainwater, which is freshwater, needs to be stored to the capacity in rubber dam. The perennial flows of the river shall continue to be passed down the rubber dam as usual, and as such there shall be no adverse effect on the morphology of the river and environment.

However, model studies are prescribed to be taken up to assess the regime status of the river in the vicinity of the rubber dam specially in alluvium stretches of river to arrive at the water way; model studies shall also be resorted to study the effect of rubber dam on the behavior of the river both upstream and downstream keeping on view the structure across the river and lines of communication both upstream and downstream.

2.7 Rubber Dam Storage

2.7.1 Capacity and Filling Time

With the cross sections of rivers and design of rubber dam cross section the gross storage of water at the proposed rubber dam needs to be computed. Based on the river cross section and longitudinal section the cross-sectional area at different pond levels may be calculated for finding out total volume.

2.7.2 Seepage Losses From the Pond

Seepage losses are to be worked out as conveyance losses. Wetland perimeter of the pond needs to be calculated to find out the seepage loss from the given period. The seepage loss values suggested by Etchevery and Harding (1933) are shown in Table 2.1. These are originally given as cubic feet/square feet of wetted perimeter/24 hours and have been converted into cumecs/million square meters of wetted perimeter.

Table 2.1 **Conveyance Losses (Etchevery and Harding, 1933)**

	Losses	
Character of Material	**Cubic feet/square feet of Wetted Perimeter in 24 h**	**Cumecs/million square meter of Wetted Perimeter**
Impervious clay loam	0.25	0.92–1.20
Medium clay loam under laid with hard pan at depth of not over 0.60–0.90 m below bed	0.35–0.50	1.2–1.8
Ordinary clay loam, silty soil, or lava ash loam	0.50–0.75	1.8–2.7
Gravelly or sandy clay loam, cemented gravel, sand, and clay	0.75–1.00	2.7–3.6
Sandy loam	1.00–1.50	3.6–5.2
Loose sandy soils	1.50–1.75	5.2–6.1
Gravelly sandy soils	2.00–2.50	7.0–8.8
Porous gravelly soils	2.50–4.00	8.8–10.6
Very gravelly soils	3.50–6.00	10.6–21.2

2.7.3 Evaporation Losses From the Pond

Losses due to evaporation will also take place and are to be completed. This may be remembered that during this period, evaporation as well as seepage losses will also be there by the river flow itself and hence may be accordingly adjusted.

2.8 Typical Installations of Rubber Dam for Water Conservation

2.8.1 Linyi City of Shandong State in China (Irrigation and Water Resources Department, 2015b)

Linyi city, the largest city of the Shandong province in China with 3 million population, with Yihe river passing through the city, has nine water-filled rubber dams constructed in cascade with the primary objectives of water conservation and river landscape improvement. These rubber dams in cascade was constructed purely for water conservation and eco-friendly waterfront revitalization and created a high level of public and political acceptance. A large number of citizens use river fronts for the recreation and fishing. The world's largest rubber dam is Xiaobudong on river Yihe, Linyi city, Shandong province and is listed in Guinness Book of Records (length 1120 m, height 3.5 m). The layout of these dams has been shown in Fig. 2.1.

The river Yihe has a general slope of 0.5 m/km (1:2000) and flood discharge of 15,000 m^3/s (5.3 lakhs cusecs) for a return period of 50 years. All the dams on river are built from time to time starting from 1996. The Xiaobudong rubber dam is a very high attraction among residents and tourists. The dam also has five under sluice steel gates to pass flash floods in addition to rubber dam. The dam was constructed in 1-year period during 1997. Thereafter, only rubber membrane was replaced in 2011. The dam has only US$150,000 per year annual operation and maintenance (O&M) expenditure including tourist handling. Some pictures of these rubber dams are shown in Fig. 2.2. The details of some of these rubber dams are given in Table 2.2.

2.8.2 Nakdonggang River in the Daegu City of South Korea (Irrigation and Water Resources Department, 2015b)

Nakdonggang River in the city of Daegu city of South Korea has nine rubber dams from 0.7 to 3 m high and 60 m width in series on the

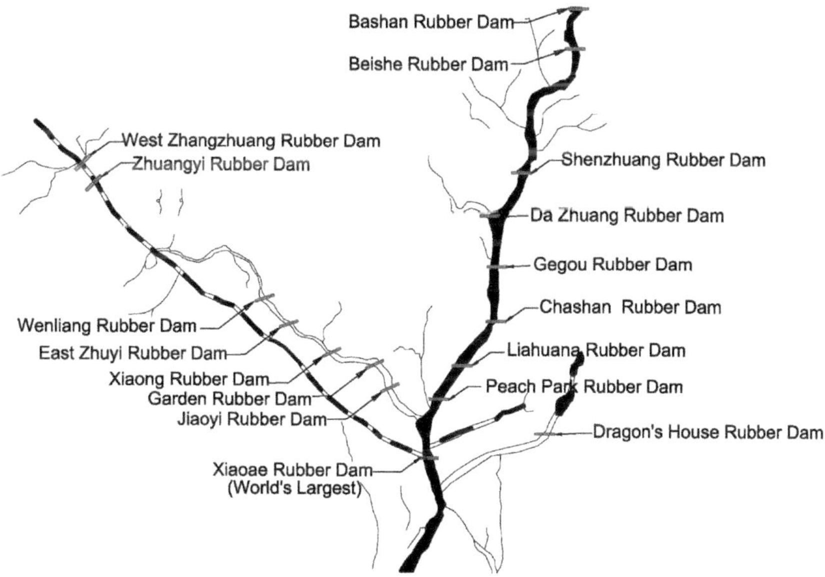

Figure 2.1 Layout of rubber dams in cascade on river Yihe in China.

Taoyuan L = 770 m x 4.5 m H with navigation bay on River Yie

Xiaobudong rubber dam on river Yie L = 1120 m x H = 3.5 m world's largest rubber dam flood discharge 15,000 m^3/s (0.53 million ft^3/s)

Figure 2.2 Rubber dams on river Yihe in Shandong Province, China.

Table 2.2 Typical Rubber Dam in Shandong Province on River Yihe in China

Sl. No.	Site	Length (m)	Height (m)	Installation (year)	End Use	Flood Discharge (m^3/s)	Remarks
1.	Taoyuan, river Yihe, Linyi	770 (10 spans)	4.5		• Navigation • Water conservation • Landscape	15,000 m^3/s (5.3 lakhs cusecs)	Water filled
2.	Xiaobudong, river Yihe, Linyi	1120 (16 spans)	3.5	1997/2011	• Water conservation • Landscape • Hydropower	15,000 m^3/s (5.3 lakhs cusecs)	Water world's largest rubber dam
3.	River Yihe, Linyi	500 (6 spans)	5	2014	• Water conservation • Hydropower		Newly constructed

Table 2.3 Examples of Rubber Dams in India

Year of Construction	Height (m)	Length (m)	Name of Site/Place	Mean Sea Level (m)	River	Purpose
2008	1.35	80.00	21.1 km Reduced Distance (RD)—High Court Hyderabad	150	Musi river	Conservation
2008	1.30	73.00	22.1 km RD—Salar Jung Museum Hyderabad	150	Musi river	Conservation

city drain/river for environmental/eco-friendly waterfront revitalization. Each rubber dam had aerators in systematic way for improving esthetic and water quality.

2.8.3 Musi River in the City of Hyderabad, India (Islam and Kumar, 2016)

Two rubber dams have been installed in the city of Hyderabad, India for water conservation and landscape. The details and photographs are given in Table 2.3 and Fig. 2.3 respectively.

2.8.4 Artificial Recharge of Groundwater From Rubber Dams (Markus, 2000)

Orange County, California, where natural recharge is insufficient to satisfy water demands, has installed rubber dams to recharge of groundwater. The area is part of a semiarid coastal plain with fewer than 50 cm of rainfall each year and is supplemented with artificial recharge to supply 70% of the water needs of its more than 2 million residents. Under Santa Ana River Inflatable Rubber Dam Project, two inflatable rubber dams with bypass facilities to divert river flows to off-river recharge basins have been constructed. These dams, both of which are 2.1 m in height and stretch 97.5 m across the Santa Ana River, are among the largest single span systems of this type in North America. They allow the district to recharge as much of the river flows as practically possible by capturing excess flows that would otherwise be lost to the ocean.

Near S.J. Museum on Musi river, Hyderabad

Near High Court on Musi river, Hyderabad

Figure 2.3 Rubber dams in India.

2.9 Fluid Adopted for Rubber Dam

Water or Air is Used for Inflating the Rubber Dam

Aspect	Air Filled	Water Filled
Operation time	Very quick	Slow
Power consumption	Low	Medium
Source of filling media	Normal air from environment	High-quality silt free (either from river or from groundwater with storage)
Stability during operation	Thick membrane required for stability	Relatively stable
Water depth regulation	Possible	Possible

In the case of rivers, water quality during lean season is highly polluted and thereafter polluted water shall not be used for rubber dam filling. In case the water-filled rubber dam is used in such situation, the water for filling has to be stored by extracting from the groundwater or after treating the water from river. This will require additional works, space, and costs.

2.10 Aerators for Improving Water Quality (Irrigation and Water Resources Department, 2015a)

As the water of many rivers passing through urban area specially during lean season has high contents of BOD and low contents of DO, it is proposed to aerate the water body created. The aeration may be done with slow movement to have an effective enrichment of water with oxygen. The floating aerators can be operated using external electricity or solar-powered electricity. The typical sections for these aerators are given as Figs. 2.4 and 2.5.

During flood the aerators will be brought to the bank and properly anchored. In view of the power situation in the state and difficulty in laying down the cables in the water body, it is proposed that floating

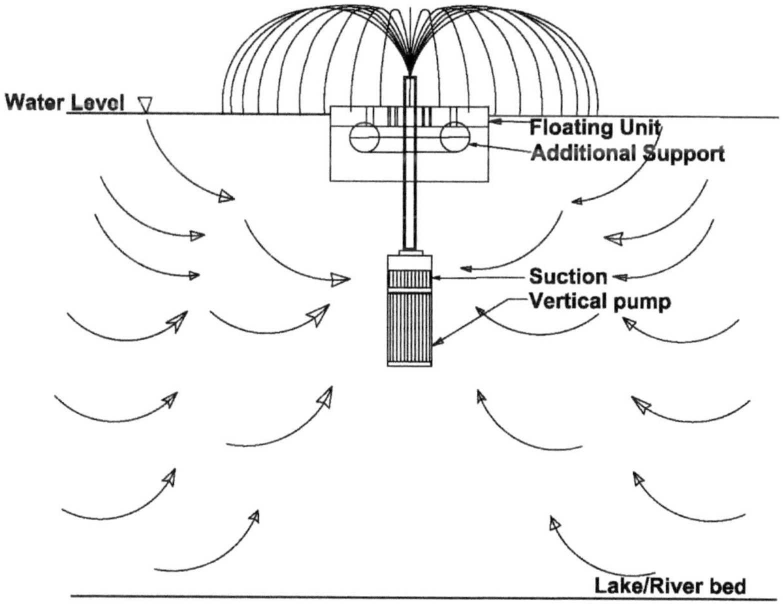

Figure 2.4 Electric-operated aerators.

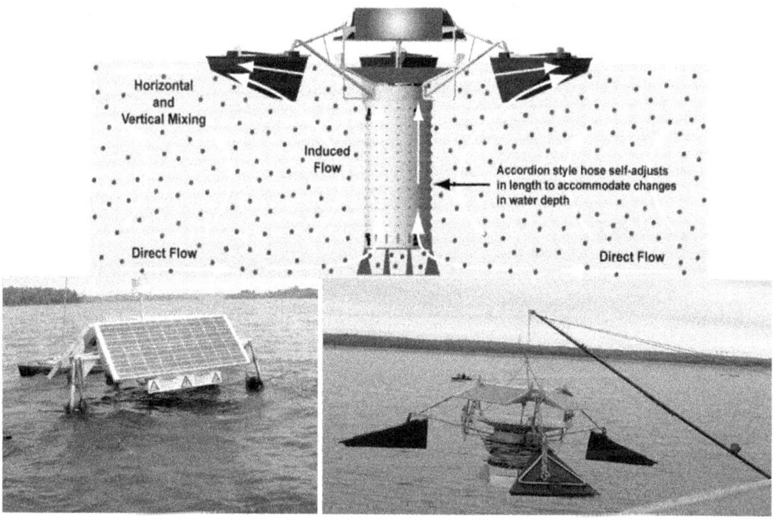

Figure 2.5 Solar-operated aerators.

solar-powered circulation equipment for wastewater treatment and freshwater application is installed in the pond created by the rubber dam. Some of the benefits of solar-based aerator are summarized as follows:

1. To control blue-green algae (cyanobacteria) through radial long-distance circulation.
2. To control invasive aquatic weeds.
3. To control inorganic chemical release from river sediments.
4. To control impaired fisheries low DO levels.
5. To increase the DO in river and hence reduce BOD in river.

2.11 Ecological Manifestation

Aquatic ecosystem supports the growth of diverse flora and fauna which varies from tiny microscopic plankton species to large fish and some higher aquatic vertebrates. Water quality also strongly influences the distribution and extent of biodiversity. These organisms are critically linked to changes in ambient environment and determine the ecological structure and functions. The producers are represented by algal group, while consumers include protozoa, rotifers, crustaceans, and fish. The decomposers are represented by a milieu of bacteria and fungus.

The stretch of some rivers based on hydrological and ecological information is referred as Eutrophic zone. This zone has slow moving waters with copious growth of algae supported by rich nutrients. Algae are food to other forms, and a food web is established ending at fish.

2.11.1 Producers (Phytoplankton)

The algae in this zone are largely planktonic. Periphyton is conspicuous by their absence. The diatoms (Bacillariophyceae), green algae (Chlorophyceae), and blue-green algae (Cyanophyceae) are the major classes of algae.

2.11.2 Consumers (Zooplankton, Etc.)

The zooplankton are represented by Protozoa, Rotifera, and Crustacea. Zoobenthos is represented by insect larvae, few annelids, molluscs, and nematodes. A large variety of fish which feed on algae, zooplankton, and zoobenthos have also been reported. The rubber dam proposed will hold water during lean season and support algal production and provide food for fish, special carps. They grow well in such conditions. Most of these fishes breed during some months; May–August would find conditions supporting broods.

2.11.3 Various Animals and Other

Fish production and catch goes down considerably in rivers having large extraction and very low lean flows along with high-polluted discharge. The increase in water storage in the area will enhance growth of fish (similar to culture in ponds/pools). If the fishing in these areas of river is regularized, production/catch will go up easily.

2.12 Impact on Environment and Biodiversity

The construction of rubber dam will have temporary impact during construction as it should involve excavation, concreting, and installation of rubber dam, etc. The impact shall be in air which shall be mitigated by adopting standard method of water sprinkling and noise barriers. There shall be positive impact of rubber dam on groundwater due to creation of water body. There shall be also positive impact of moisture and temperature due to creation of water body. Due to creation of water body, birds will be attracted to the rubber dam and will not only increase esthetic but also biodiversity.

2.13 Impact on Society

Creation of water body will enhance not only the water quality but also improve the general esthetic and landscape of the area which otherwise remains dry for several months giving ugly picture. The farmers who normally use the river cross section in summer for fruits/vegetables will be marginally impacted even though the pond-submerged area will be within the main river cross section.

References

CWC, 2005. General Guidelines for Water Audit & Water Conservation. Central Water Commission, Evaluation of Water Utilization Directorate, Ministry of Water Resources, p. 52 <http://www.cwc.gov.in/main/downloads/Water%20Audit%20&%20Water%20Conservation%20Final.pdf>.

Etchevery, B.A., Harding, S.T., 1933. second ed. Irrigation Practice and Engineering, vol. II. McGraw Hill Book Co. Inc., New York, p. 106.

Islam, S.U., Kumar, A., 2016. Inflatable dams for SHP projects. Renew. Sustain. Energy Rev. 57, 945–952.

Markus, M.R., 2000. Aquifer recharge enhanced with rubber dam installations. Roads & Bridges, December 28, 2000. <https://www.roadsbridges.com/aquifer-recharge-enhanced-ruber-dam-installations>.

Irrigation and Water Resources Department, 2015a. Project Report, Project for Constructing Water Augmentation Structure on River Yamuna Near Vrindavan. Irrigation & Water Resources Department, Uttar Pradesh (unpublished).

Irrigation and Water Resources Department, 2015b. Site Visit Report, Rubber Dam Installations and Manufacturing Facilities in China, South Korea and Austria. Irrigation and Water Resources Department, Govt. of UP and AHEC IIT, Roorkee (unpublished).

3 Raw Materials—Rubber Dam

Abitha V.K. [1], ***Sabu Thomas*** [1,2], ***Ajay Vasudeo Rane*** [3] ***and Krishnan Kanny*** [3]

[1]School of Chemical Sciences, Mahatma Gandhi University, Kottayam, India, [2]International and Interuniversity Centre for Nanoscience and Nanotechnology, Mahatma Gandhi University, Kottayam, India, [3]Composite Research Group, Department of Mechanical Engineering, Durban University of Technology, Durban, South Africa

Abstract

Rubbers form a special class of materials characterized by a unique combination of useful properties such as elasticity, flexibility, toughness, and impermeability. The uses of rubbers include both engineering and speciality applications. The properties of rubber products depend not only on the characteristics of rubbers but also on the various additives and ingredients mixed into the basic rubber to form a rubber compound. The selection of additives and their incorporation into the rubber to improve the properties of a raw rubber is based more on experience and art than on a rational or scientific approach. The rubber industry has innovated and achieved a lot over the last few years. Service lives of rubber products have dramatically improved. None of these innovations would have been possible without an emphasis on understanding the chemistry of raw materials and compounds. Generally a tubular composites assembly is divided into an outer layer, middle layer, and an inner layer serving the function of protection, adhesion, and impermeability to the final structure. Modification in engineering aspects, that is, gauge size (thickness), has an influence on the service of the fabricated composites, but as this chapter is dedicated to raw materials for rubber dam, information on raw materials used for achieving specified properties and basic rubber compounding is described.

Keywords: Rubber; filler; specialty additives; raw materials; textile fabric

3.1 Introduction

Rubbers form a special class of materials characterized by a unique combination of useful properties such as elasticity, flexibility, toughness,

Hydraulic Rubber Dam. DOI: https://doi.org/10.1016/B978-0-12-812210-5.00003-1

and impermeability. The uses of rubbers include both engineering and speciality applications. The properties of rubber products depend not only on the characteristics of rubbers but also on the various additives and ingredients mixed into the basic rubber to form a rubber compound. The selection of additives and their incorporation into the rubber to improve the properties of a raw rubber is based more on experience and art than on a rational or scientific approach (Franta, 1989). The rubber industry has innovated and achieved a lot over the last few years. Service lives of rubber products have dramatically improved. None of these innovations would have been possible without an emphasis on understanding the chemistry of raw materials and compounds. Generally a tubular composites assembly is divided into an outer layer, middle layer, and an inner layer serving the function of protection, adhesion, and impermeability to the final structure. Modification in engineering aspects, that is, gauge size (thickness), has an influence on the service of the fabricated composites, but as this chapter is dedicated to raw materials for rubber dam, information on raw materials used for achieving specified properties and basic rubber compounding is described.

As the name suggests, "Rubber Dam," the most important raw material comes into our mind is rubber. Rubber dam also in coined as textile reinforced rubber dam; hence, rubber and textile are the two prominent raw materials used in fabrication of rubber dam (fabrication techniques for rubber dams are mentioned in Chapter 4: Manufacturing Process—Reinforced Rubber Sheet for Rubber Dam). Rubber provides elasticity and fiber materials possess rigidity and stiffness—combining these two materials in a technical forming method lead to a rubber sheet reinforced with fiber, which can be further used for construction of body for rubber dam. Rubbers are basically classified based on their scope of applications into general purpose rubbers and speciality rubbers. Rubbers like natural rubber (NR), styrene butadiene rubber (SBR), and butadiene rubber are classified under general purpose rubbers. Speciality rubbers like ethylene propylene diene rubber, chloroprene rubber (CR), chlorosulfonated polyethylene, nitrile butadiene rubber and its hydrogenated form, acrylic rubbers, silicone, polyether, fluoroelastomers are specialty elastomers and have a variety of components and structures that have been selected to provide key performance attributes. Enhancement in physical properties and degradation resistance is required for reinforced rubber sheet for rubber dam, which is done by selecting the specific rubber and additives. Reinforced rubber sheet for rubber dam should have specific tensile strength, tensile elongation, hardness, and modulus, reducing set under compression and tension, tear resistance, low-temperature properties (for dams operated in cold regions), decreasing air permeability, rubber to fabric adhesion, flame retardancy. Apart from the physical properties, degradation properties like hot air and heat aging resistance, ozone resistance, flex and fatigue resistance, cutting and chip resistance, abrasion and/or wear resistance, weathering resistance, hydrolysis resistance also are important in order to design a reinforced rubber sheet for rubber dam. Further part of the chapter will describe the rubbers and additives which can be used in order to have an improvement in physical and degradation properties of reinforced rubber sheet for rubber dam body. As mentioned above, rubber formulations are complex systems and hence each ingredient in formulation plays an important role in imparting end properties to the fabricated product. Fig. 3.1 gives broad classification of the raw materials used in preparation of typical uncured rubber compound mixed in an internal or an open mixing mill (as discussed in Chapter 4: Manufacturing Process—Reinforced Rubber Sheet for Rubber Dam), and Fig. 3.2 outlines the general advantages of compounding rubber.

Rubber/textile composites are majority of polymer/textile composites on which the modern world relies for transportation, material handling systems, mechanical drive system in the form of textile/rubber tyres, textile/rubber conveyor belting, and textile/rubber drive belts, respectively. Miscellaneous applications of textile/rubber composites include hoses, hovercraft skirts, air brake chamber diaphragms, snow mobile tracks, inflatable structures (inflatable boats, oil booms, dunnage bags,

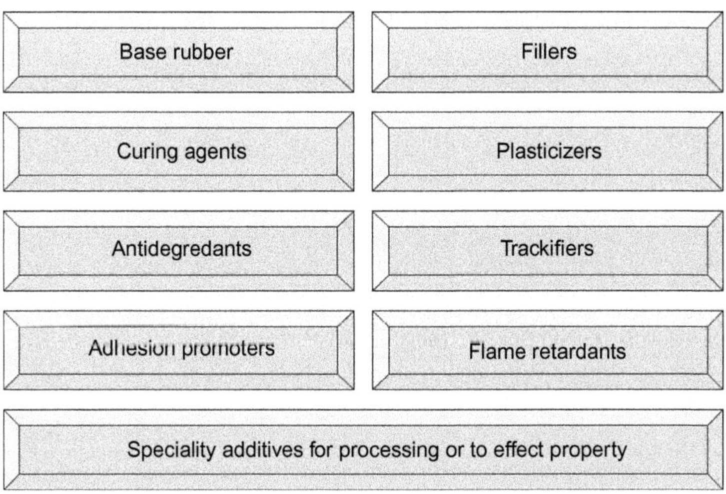

Figure 3.1 Raw materials for the preparation of uncured rubber compound via internal or open mill mixing technique.

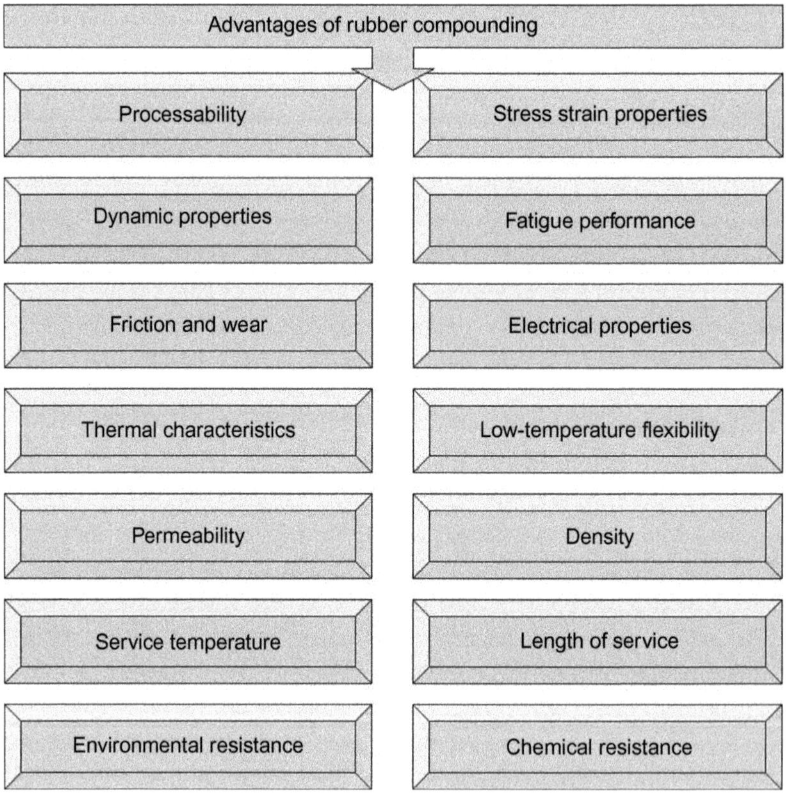

Figure 3.2 General advantages obtained by compounding.

inflatable buildings, and inflatable dams), and noninflatable structures (flexible storage tanks, reservoirs liners, pond liners). As per reports, usage of rubber/textiles together started longtime after invention of textile and rubber. Owing to the technological advances in material science, replacement of natural materials by synthetic materials took place, both for polymers and fibrous reinforcement leading to a concept of tailor composites, with specific properties and required performance. Adhesion of textiles with rubber was a major problem, hence with each new synthetic polymer; adhesive systems were developed in order to achieve optimum performance from textile/polymer composites (Wootton, 2001).

As we know, rubber dam is mainly made up of rubber and fabric. Let us first discuss about rubber, when we mean rubber that is, rubber compound. The rubber compound is made by mixing raw rubber, reinforcement material, softener, and various other chemicals as mentioned in Fig. 3.1. Different properties are required for components of rubber dam, depending on application. Outer layer of the rubber dam should possess heat resistance, wear, and cut resistance along with ultraviolet resistance, weather resistance, aging resistance, oxygen and ozone resistance, and excellent mechanical properties (tensile strength and modulus); in fact outer layer of rubber dam protects the entire rubber body from environmental factors and hence should be compounded carefully to achieve said properties. Inner layer of the rubber dam should possess specific mechanical properties (tensile strength and modulus), but importantly air retention and water retention are the primary function of the inner rubber compound. Middle layer composes of fabric coated with rubber (via calendering process—described in Chapter 4: Manufacturing Process—Reinforced Rubber Sheet for Rubber Dam) serves the function of providing strength to the rubber dam structure, also it adheres the outer and inner layer and makes the structure of rubber body more strong. In fact construction of this layer determines the strength of rubber dam. Now let us discuss about the rubbers and their types that performs very important functions. The rubbers used industrially for fabrication of reinforced rubber sheet are listed down 1–7, which includes general purpose and speciality rubbers. Selection of a base rubber has the greatest effect on the properties of a rubber compound and final product. There are variety of grades for one rubber, and hence using different grades will have an impact on property; hence, many rubber formulations are based on blends of two or more rubbers, that is, NR/SBR, NR/BR and SBR/BR and many more possible combinations (depending on their miscibility with each other) can be formulated (Dick, 2014). From the list of rubbers mentioned below—NR, SBR, blends of NR and SBR, polybutadiene rubber (PBR) (in blend with NR/SBR) ethylene propylene diene monomer (EPDM) on specific compounding are suitable rubbers for fabricating rubber dam body. Considering properties of CR and chlorosulfonated polyethylene rubber (CSM), they can be used as an outer protective layer for rubber dam body. Owing to excellent low-permeability properties, butyl rubber (IIR) compound shall be used as an inner liner in rubber body.

1. NR
2. SBR
3. PBR
4. EPDM
5. CR
6. CSM
7. IIR

Table 3.1 and Fig. 3.3 gives overall properties and their chemical structures respectively, for rubbers listed above. Polyisoprene rubber is

Table 3.1 Properties of Rubbers (Herco Rub Technical Article)

NR	SBR	PBR
Good dynamic properties, tensile and tear strength, abrasive resistance, poor resistance to chemicals and oil substances, no longtime exposure to sunlight, ozone, and heat	Good tensile strength, dynamic properties and abrasive resistance, poor resistance to chemicals and oil substances, no longtime exposure to sunlight, ozone, and heat	Excellent dynamic properties, good cold resistance, good tear strength, poor resistance to chemicals and oil substances No longtime exposure to sunlight, ozone, and heat
EPDM	**CR**	**CSM**
Excellent resistance to ozone, oxidants and weather (water) conditions, good heat resistance, excellent insulator, good chemical resistance (not oil), poor resistance to oil substances, less mechanical properties compared to NR	Good resistance to ozone, weather conditions (water) and sunlight, good chemical and medium oil resistance, good high temperature resistance, expensive than general purpose synthetic rubbers, no resistance to strong oxidizing acids, esters, ketones, chlorinated—and aromatic hydrocarbons	Good heat resistance, good ozone and weather (water) resistance, good chemical resistance, color-proof, excellent abrasive resistance, intermediate oil resistance
IIR		
Excellent resistance to gas permeability, good resistance to chemical, ozone, weather, and oil, intermediate mechanical properties		

SBR, Styrene butadiene rubber; *NR*, natural rubber; *PBR*, polybutadiene rubber; *EPDM*, ethylene propylene diene monomer; *CR*, chloroprene rubber; *CSM*, chlorosulfonated polyethylene rubber; *IIR*, butyl rubber.

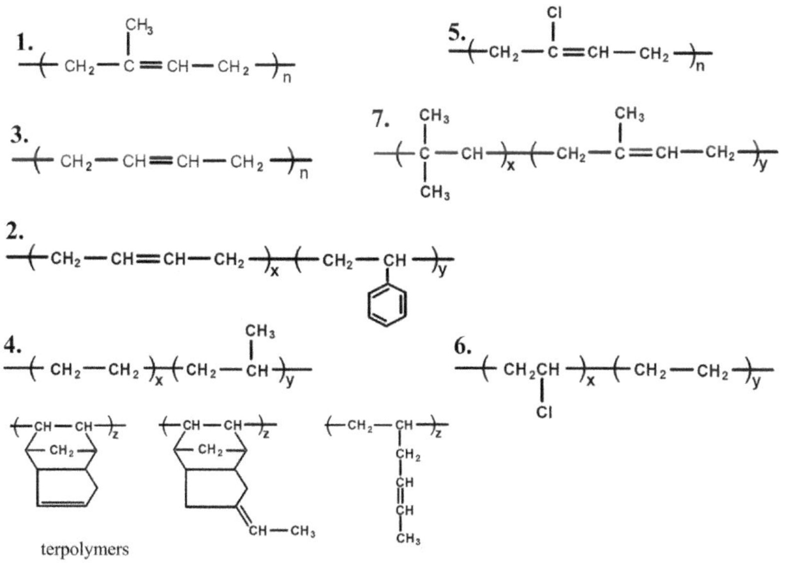

Figure 3.3 Chemical structure of rubbers (numbered as per list).

produced by artificially synthesizing isoprene which is a principle constituent of NR. Its characteristics are naturally quite similar to those of NR. The main difference between the two is quality; polyisoprene rubber is more consistent than NR since it does not contain natural impurities. The downside however is the cost is higher than NR.

NR due to its high structural regularity tends to crystallize at low temperature or when it is stretched. Low-temperature crystallization causes stiffening but can be easily reversed by warming. Strain-induced crystallization gives NR high tensile strength and resistance to cutting tearing and abrasion resistance (Ciullo and Hewitt, 1999). Hot emulsion SBR are difficult to mill, mix, and calender than NR; they are deficient in tack and have relatively poor inherent physical properties, addition of process oil and reinforcing fillers improves processability, and physical properties of hot emulsion SBR. Cold SBR generally has higher molecular weight and narrow molecular weight distribution; hence, it offers better abrasion and wear resistance in addition to greater tensile strength and modulus than hot SBR, since high molecular weight makes it difficult to process it is supplied in oil extended form. Solution SBR are comparable in tensile, modulus, and elongation but offers lower heat build-up, increase in flex and resilience. Solution BR with greater than 90% cis 1,4-polybutadiene configuration hardens at temperature lower than NR and other commercial rubbers, this gives low-temperature flexibility, greater resilience, less heat build-up under continuous dynamic deformation. cis-BR possess superior abrasion resistance, and great tolerance to accept fillers and plasticizers. High cis-BR are blended with SBR and NR to improve processing, and BR imparting its desirable properties to the blend. BR blends with NR and SBR improves resistance to heat build-up, abrasion resistance, and cut growth resistance. High cis-BR and medium cis-BR have excellent abrasion resistance, high vinyl BR downsides abrasion resistance (Ciullo and Hewitt, 1999). BR improves durability, abrasion, and flex crack resistance in rubber blends. IIR is saturated rubber with repellence for water and polar liquids, and affinity toward aliphatic and some cyclic hydrocarbons. Likewise IIR is highly resistant to diffusion or solution of gaseous molecules; hence, air permeability is primary property for commercialization. The saturated backbone imparts ozone resistance, also makes the rubber less flexible and delayed elastic response to deformation than other rubbers, this property imparts vibration damping and shock absorption properties to IIR. IIR possess low gas and moisture permeability, resistance to aging and weathering, chemical and biological inertness, good self-sealing, and low fragmentation during needle penetration (Ciullo and Hewitt, 1999). CR due to its high structural regularity (trans) possess strain-induced crystallization that results in higher tensile strength as for NR. CR are cured with oxides of lead, magnesium, and zinc which in-turn enhances water resistance, chlorine atom present in CR imparts resistance to oil, weathering, ozone and heat, activity of double bond is decreased due to presence of chlorine, thereby increasing oxidative resistance. Low-temperature property and price makes CR inferior to NR. EPDM are compounded to provide good low-temperature flexibility, high tensile strength, high tear and abrasion resistance, excellent weather ability, good electrical properties, and high heat resistance. High molecular weight EPDM can incorporate high level of fillers. EPDM have low resistance to hydrocarbon oils, lack of building tack is compensated by the use of resins. CSM compounds are odorless, light stable, and easily colored. CSM is highly resistant to oxygen, ozone, weathering, corrosive chemicals, abrasion, low temperature, oil, and grease. CSM possess high tensile properties without incorporating high reinforcing fillers (Ciullo and Hewitt, 1999).

Variety of ingredients like accelerators, activators, antidegradants, antireversion agents, antistatic agents, biocides and fungicides, bonding additives, factice, fillers, fire retardants, peroxides, petroleum oils, pigments, plasticizers, prevulcanization inhibitors, release agents, resins, vulcanizing agents, waxes, blowing agents, deodorants and reodorants, desiccants, dusting and antitack agents, process additives are used in rubber industry as per their requirement in a rubber formulation (Richard, 2002). Different rubbers are formulated as per specifications. As per literature, tensile strength, elongation at break, permanent deformation at break, hardness, brittleness temperature, hot water aging (100°C × 96 hours) and (70°C × 96 hours), volume expansion ratio, ozone aging (10,000 pphm, temperature 40°C, stretching 20%, no cracking), wear loss, flexibility are the specific properties for rubber dam. Tensile strength of ~14, ~12, and ~6 MPa, elongation at break of ~400%, ~400%, and ~250%, permanent deformation of ~30%, ~30%, and ~35%, hardness of range ~55–65, ~50–60, and ~55–65, brittleness temperature of −30°C, ozone aging with no cracks till 120 minutes, wear loss of not more than 1.0 cm^3/1.6 km and flexibility of not less than 20,000 cycles with no cracks are required for outer, middle, and inner, and cover rubber compounds (http://www.inflatabledam.com/inflatabledam/air-inflatable-rubber-dam.html).

Cotton, rayon, nylon, polyester, and aramid are the five main types of fibers used in fabrication of textile/rubber composites. Cotton (cellulose), naturally occurring fiber, and is used till now in many applications but due to advances in material sciences is being replaced by manmade fibers. Rayon (regenerated cellulose), very similar to cotton chemically but the process in which yarn is produced, differs its properties with rayon. Cotton processes itself through a series of processes before it forms a strong yarn—bale breaking, opening and cleaning, carding, drafting, spinning; whereas during production of rayon, cotton is boiled in caustic soda, forming soda cellulose, which further forms sodium cellulose xanthate by treating with carbon disulfide, and lastly forms rayon after treatment with sulfuric acid. Nylon 6 and Nylon 66 (linear aliphatic polyamides) offer advantages over cellulosic fibers and hence was accepted by rubber industries (Wootton, 2001). Nylon 6 is polymerized product of caprolactam, which has six carbon atoms, hence the name, whereas polycondensation of hexamethylenediamine (6 C atoms) and adipic acid (6 C atoms) forms Nylon 66. Further polyester also possess strength equivalent to nylons, but with high modulus values which made suitable for speciality applications. Polyesters are manufactured via ester interchange method (terephthalic acid—dimethyl terephthalate—polyester) and direct esterification method (terephthalic acid + ethylene glycol). Aramid fibers possess high strength and modulus but are limited to use considering the high cost in comparison to its counter parts. Aramids (aromatic polyamides) where replacement of aliphatic backbone by aromatic groups brings changes in properties of fibers, in comparison to Nylon 6 and Nylon 66. Table 3.2 gives

Table 3.2 General Physical Properties of Textile Fibers (Wootton, 2001)

	Specific Gravity	Tensile Strength (MPa)	Elongation at Break (%)	Initial Modulus (cN/Tex)	Mean Filament Diameter (μm)
Cotton	1.54	230	8	225	15
Viscose rayon	1.52	685	10	600	8
Polynosic rayon	1.52	850	6	800	8
Nylon 6	1.14	850	19	300	25
Nylon 66	1.14	950	16	500	25
Polyester	1.38	1100	13	850	25
Aramid	1.44	2750	4	4000	12

information of general physical properties of five fibers used in rubber industry (Wootton, 2001).

Selection of fabric type and construction depends on the property for which textile/rubber composites is designed, that is, strength—considering working load for specific application. Properties such as elongation, flexibility, fatigue resistance, tear strength, gauge are also important and should be inline as specified for applications. User should specify the properties so that designing of fabric can be done to achieve specific reinforcement. Generally, synthetic fibers provide higher strength with lower fabric weight, but for example if bulk is required, two to three plies of cotton fabric serve the purpose. Finally after deciding through the general physical properties of textile fibers, type of fiber, the fabric structure design, that is, construction must be considered (Wootton, 2001). Table 3.3 briefs about the different construction.

One of the most important materials produced by the silicate industry is cement, employed in enormous quantities in all kinds of construction work. As it is inert toward water, it is absolutely irreplaceable in all marine structures, for construction of dams, and buildings on damp places. Mortar mixed with gravel, crushed rock, etc., is known as concrete. Concrete is used to build different structures like bridges, vaults, cisterns, tanks etc. Structures made of concrete based on iron beams and rods are called reinforced concrete (Hajra Choudhury, 1977). Reinforced concrete foundation is required to fix rubber sheet in order to complete final structure of a rubber dam. Reinforced concrete is made by casting concrete over steel rods. Like any brittle material,

Table 3.3 Construction in Fabric Structure (Wootton, 2001)

Construction	Features
1 Plain weave	• Basic woven structure and widely used • Coating application
2 Matt weave	• Suitable for tensile strength improvement • Increases tear strength
3 Twill	• Not suitable to mechanical fastening operations, as this operation can comb out the woven structure • Inclusion of crow feet gives higher comb out resistance and improves mechanical fastening retention • Proofings, hose, hovercraft skirts, plied beltings
4 Leno weave	• Fabrics possess good tear strength • Used as breakers (additional layer of fabric reinforcement close to surface of final composite, to impart improved tear and cut resistance to final composite)
5 Stress warp fabric structures	• Very high-strength fabrics • Single- or two-ply reinforcement of belting • Reducing binder reduces mechanical fastening retention of final fabric • Increasing number of binders increases the fastening retention property of final composite
6 Solid woven fabric structures	• Double equal fabric • Carcass of heavy duty solid woven PVC conveyor belting
7 Triaxial woven fabric structures	• Properties uniform in all directions • Much improvement in tear strength

PVC, Poly (vinyl chloride).

concrete is much stronger in compression than in tension. Placing right steel bars of the right size and shape in concrete mix and by strategically locating them in the cross-section of the structural element, particularly with respect to the applied load, it is possible to utilize full compressive strength of the concrete and to compensate for its weakness in tension by the full resistance of the supporting load. Concrete may also be prestressed so that it can support a tensile load, by putting steel rods under tension and pouring the concrete over them (Hajra Choudhury, 1977). After the concrete has set and hardened, the tensile load is released from the rods, thus placing it in compression. A tensile load must exceed this compressive strength before a net tensile stress exists. Further details on fixation of rubber sheet using clamping plates, anchor bolts or by using any other mechanical fasteners are discussed in chapter related to anchoring and bonding of rubber dam.

To conclude, rubber compounding is the skill and knowledge of choosing different compounding ingredients and their measures to mix and produce a useful rubber formulation that is processable, meets or exceeds the customer's final product requirements, and can be competitively priced. In case of inflatable rubber dam, rubber compounding for outer, middle, and inner layer differs, owing to their function during service. Hence, a rubber compounder should carefully select the ingredients and its amount while designing a cost-effective rubber compound as per optimum properties requirements. Skill and knowledge leads to good compounding practice, that is, equilibrium of properties.

References

Ciullo, P.A., Hewitt, N., 1999. The Rubber Formulary. Noyes Publications, Nowrich, New York, USA.

Dick, J.S., 2014. How to Improve Rubber Compounds – 1800 Experimental Ideas for Problem Solving, second ed. Hanser Publishers, Cincinnati, OH, ISBN 978–1–56990–533-3.

Franta, I., 1989. Elastomers and Rubber Compounding Materials – Manufacture, Properties and Applications. Elsevier Publishers, Netherlands, ISBN-0-444-98906-4.

Hajra Choudhury, S.K., 1977. Materials Science and Processes. J K Publishers, ISBN: 9780906216002.

Herco Rub Technical Article. Advantages and Limitations of Some Elastomer Types.

Simpson, R.B., 2002. Rubber Basics. Rapra Technology Limited, UK, ISBN: 1-85957-307-X.

Wootton, D.B., 2001. The Application of Textiles in Rubber. Rapra Technology Limited, UK, ISBN: 1-85957-277-4.

4 Manufacturing Process—Reinforced Rubber Sheet for Rubber Dam

Ajay Vasudeo Rane and Krishnan Kanny
Composite Research Group, Department of Mechanical Engineering, Durban University of Technology, Durban, South Africa

Abstract

Manufacturing process are the steps through which raw materials are transformed into a final product. The manufacturing process begins with the product design and material specification from which the product is made. These materials are then modified through manufacturing processes to become required part. The manufacturing process of reinforced rubber sheet for rubber dam consists of some basic processes that are described in the chapter.

Keywords: Assembly; calendering; frictioning; manufacturing; rubber; rubber mixing; vulcanization

4.1 Introduction

Manufacturing process is the steps through which raw materials are transformed into a final product. The manufacturing process begins with the product design and material specification from which the product is made. These materials are then modified through manufacturing processes to become required part (https://en.wikipedia.org/wiki/Manufacturing). The manufacturing process of reinforced rubber sheet for rubber dam consists of some basic processes that are depicted in Fig. 4.1 and further described in the chapter.

Rubber compound preparation involves mixing of elastomers, fillers, and chemicals, speciality additives under shear in a mixing mill in a specific sequence as indicated in Fig. 4.2. Calendering operation on rubber compound is carried out to get rubber compound sheet of specific and uniform thickness. Processing on fabrics, that is, to coat fabrics

Hydraulic Rubber Dam. DOI: https://doi.org/10.1016/B978-0-12-812210-5.00004-3

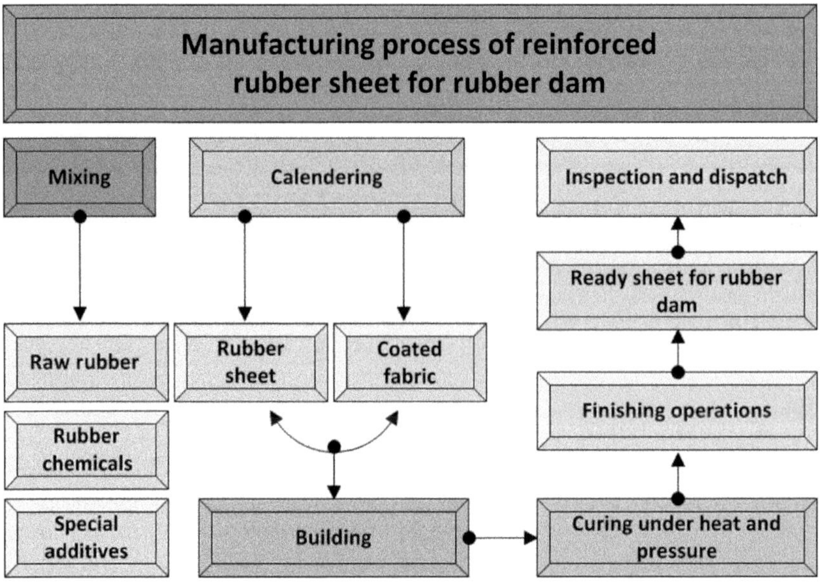

Figure 4.1 Layout for manufacturing process of a reinforced rubber sheet for rubber dam.

with rubber during calendaring operation (Erman et al., 2013), is operation that leads to fabrication of individual components. Further these components are assembled, into carcass i.e., body for rubber dam as per design criteria for application. Finally curing the assembled component using a suitable technique under heat and pressure. Quality operations involving finishing steps, inspection of final product, and dispatch specification are done. As this chapter is devoted to manufacturing process of rubber reinforced sheet for rubber dam—detailing on rubber compound preparation i.e., via internal and open mills, calendaring operations for rubber sheet and coating of fabrics, assembling of components and curing of assembled part will be discussed. Information on raw materials is dealt in Chapter 3 of the book.

Mixing raw rubbers with fillers, chemicals and special additive is first and vital operation for preparing rubber compound with uniform distribution of all compounding ingredient in a rubber compound for reinforced rubber sheet in structural application for rubber dam. Internal mixer (see Fig. 4.3) and open mills (see Fig. 4.4) are being used as equipment for mixing rubber and its additives. Mixing rubbers involves number of different mechanisms and stages, and are classified into viscosity reduction, incorporation, distribution, and dispersion (https://www.tut.fi/ms/muo/vert/8_processing/4.3.e.htm). Mastication (i.e., viscosity reduction) is first step performed in mixing process; during mixing (irrespective of

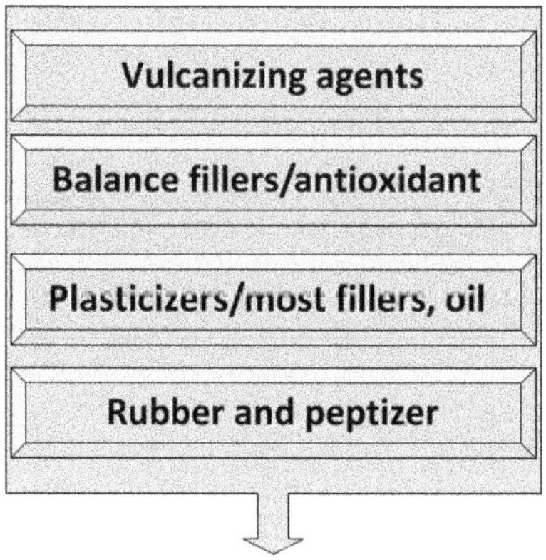

Figure 4.2 Typical mixing sequence for preparation of rubber compound.

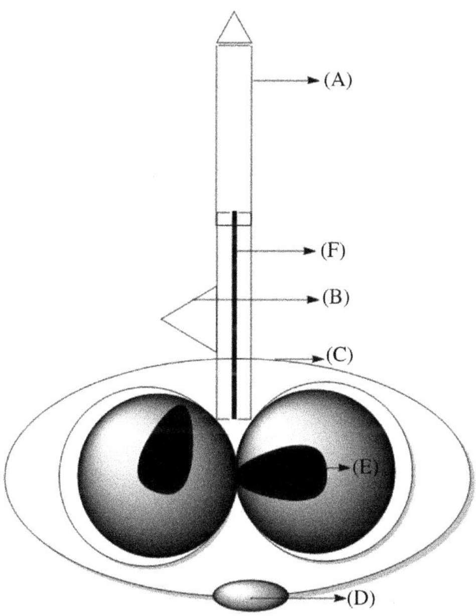

Figure 4.3 Cross-sectional view for internal mill: (A) weight cylinder, (B) feed hopper, (C) mixing chamber, (D) drop door, (E) intermeshing rotors, and (F) piston.

equipment used), rubber molecules are broken down, that is, a phase where mechanical breakdown of high molecular weight rubber occurs. Generally, molecular break down and mixing operation takes place at higher temperatures (i.e., 180°C) in an internal mixer (http://www.industrial-electronics.com/engineering-industrial/process-plant-mach_18.html). After rubber molecules are broken down, fillers, processing aids, and special additives are added to raw rubber in equipment selected for mixing

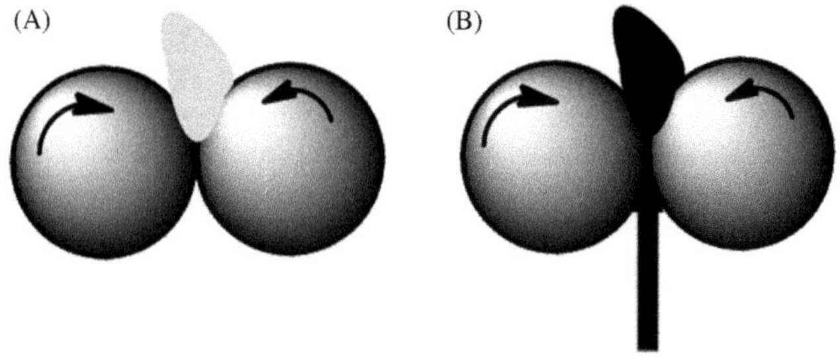

Figure 4.4 Side view of open mill (two-roll mill)—(A) raw rubber between nip gaps and (B) carbon-black-filled rubber sheeting down from nip gaps.

at specific intervals following a mixing sequence, that is, incorporation, distribution and dispersion. Even spread of filler particles into rubber matrix is termed as "distribution." Breaking up of agglomerates of filler increases the contact area between rubber matrix and filler, termed as dispersion (https://www.tut.fi/ms/muo/vert/8_processing/4.3.e.htm).

Rubber compounds are also prepared using a two-roll mill (i.e., open mill) but are less efficient in comparison to internal mixers, comparing efficiency, automation, quality and uniformity, batch weight, mixing time. Friction ratio, that is, speed of front roll to speed of back roll, clearance, rotor surface, pressure are criteria which determines the degree of breakdown of rubber molecules in a rubber mixing equipment. Typical mixing sequence is been depicted in Fig. 4.2. After mixing operation, rubber compound is sheeted and water cooled, and kept for maturation for period of 8–10 hours. Migration of additives to surface or blooming of any ingredients or any other surface deposits should be avoided (following a predetermined mixing sequence), as this interferes with adhesion of rubber compound with fabric (David, 2001).

All textile fibers increase in length, when subjected to tension stress—hence, they are stretched during heat setting which increases the modulus and reduces tendency to grow further during curing process (https://www.tut.fi/ms/muo/vert/8_processing/4.3.e.htm). As calendering operations are mostly concerned with coating of fabric, impregnation, that is treatment of fabric is an important operation before coating fabric via calendering operation in order to obtain a good adhesion between reinforcing fabric and rubber matrix, while performing curing operation. For fabrics adhesion is obtained by impregnating the fibers with adhesive solution of resorcinol–formaldehyde–latex (RFL bonding system) following dip coating technique (https://www.tut.fi/ms/muo/vert/8_processing/4.3.e.htm). The type of latex is chosen to fit the polymer in the compound to be used for coating fabric. Dip coating allows

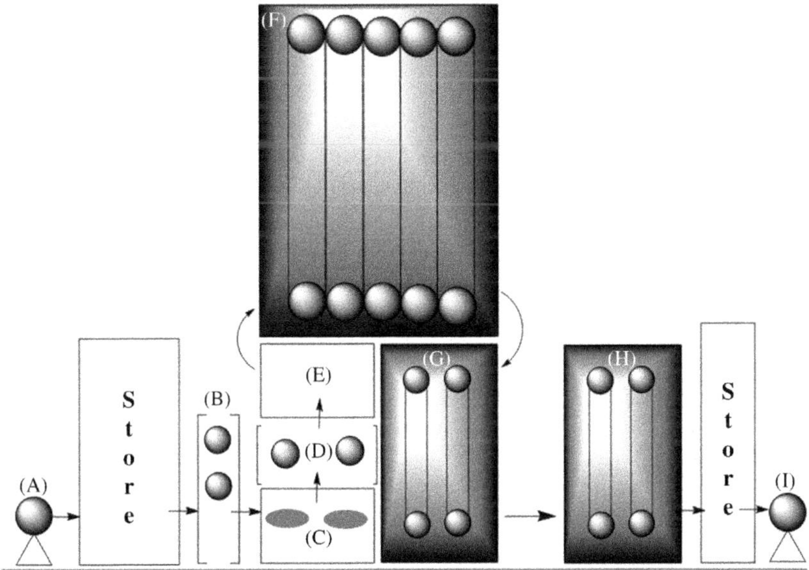

Figure 4.5 Layout of stretching unit—(A) wind off unit, (B) stretching device, (C) dip unit, (D) squeeze roll, (E) vacuum suction, (F) drying oven, (G) and (H) stretching device, and (I) wind up unit.

formation of a film on the fabric after impregnation, drying, and curing of resorcinol and formaldehyde. Film formed on fabric is chemically linked with fabric and levels the difference in module between rubber and textile (https://www.tut.fi/ms/muo/vert/8_processing/4.3.e.htm). Schematic representation of impregnation and fixation in hot stretch is represented in Fig. 4.5.

Reinforced rubber sheet is a vital part of rubber dam, surface contamination of textile or the rubber compound should be prevented in order to reduce adhesion between them. Hence calendering or spread coating is method to ensure clean contact between rubber and textile surface. Moisture in textiles is the most volatile material, and interferes with adhesion—hence cotton and rayon fabric should be dried before rubberizing or assembling. Moisture absorption for nylon and polyester is low and seldomly causes any problem, but in case of nylon at high temperatures and very high humidity, it is advisable to install a drying stage before using nylon for rubberizing (David, 2001). Coated fabric is used as reinforcement for reinforced rubber sheet used in construction of body for rubber dam. Calenders are used in rubber industry primarily to produce rubber compound webs and sheets of various thicknesses, coating textiles, or other supporting materials with thin rubber sheets or frictioning fabrics with rubber compounds (Bhowmick et al., 1994). Calender is a high-duty machine with three to four steel rolls plated with chrome, revolving in opposite direction (Erman et al., 2013). Operations in calendering machine are done by feeding the rubber compound into one of the roll gaps. Two rolls built together in a frame forms one gap and are called two-roll calender, three rolls in frame forms two gaps and is called three-roll calender, four rolls in frame forms three gaps and is called four-roll calender. Frictioning and coating are two different techniques used in calendering operation; frictioning can be carried out on three and four-roll calenders, but three-roll calender is preferred in which rubber is forced into fabric—and one-side coat is done, whereas coating is done on a four-roll calender and two-side coating is possible (https://www.tut.fi/ms/muo/vert/8_processing/4.3.e.htm). Thickness of coating depends on product it is intended for. Cooling rate for calendered sheet will influence shrinkage, and hence, fast cooling rates are preferred. Fig. 4.6 illustrates the frictioning and coating carried out using three and four-roll calenders.

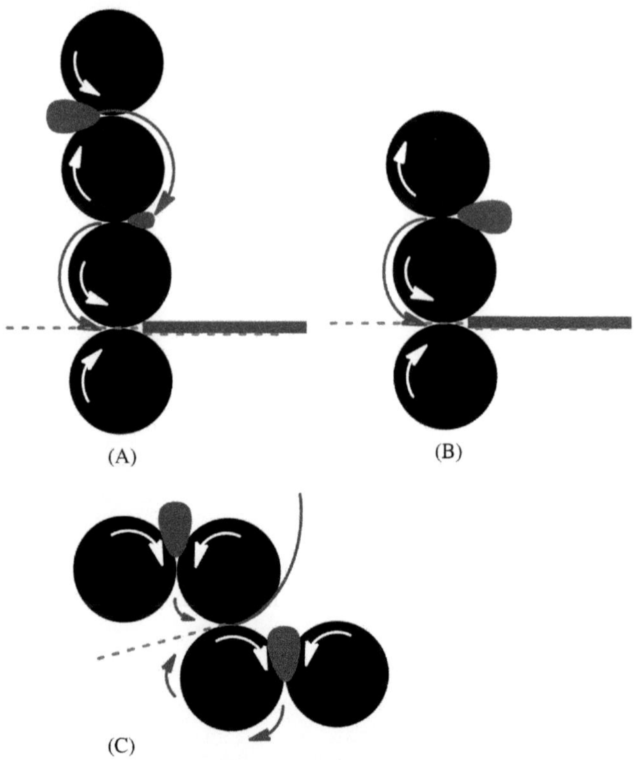

Figure 4.6 Arrangement for frictioning and coating of rubber on fabric through calendering operation (A) four-roll calender and (B) three-roll calender during frictioning (C) four-roll calender during coating.

The above process gives information on operations involved in preparation of components for reinforced rubber sheet; the latter part shall describe in short the building of green reinforced sheet and its curing process. If solvents are used as wipe to freshen the surface of the rubber, it is necessary to evaporate all the solvents before assembling, to avoid air bubbles. During assembly of components (i.e., rubber sheet and rubberized fabric), it is important to remove the entrapped air or volatile materials. Air bubbles can be eliminated by using a spiked roller or a profiled compaction roller—which egresses air out of the assembled composites (David, 2001). Minimizing the air entrapment between layers is considered as a good assembly, which improves the adhesion and in turn enhances the structural properties of final rubber dam during application. The assembly pattern (i.e., building) for reinforced rubber sheet is determined considering the application intended for. Assembly of components for rubber dams to be used as check dams in agricultural field differs from the one to be used in rivers. Manufacturing reinforced sheet is dependent on basic construction (assembly), hence process differs. In general, rubber dams used in agricultural fields as check dams can be single ply, with advantage of lower weight. Reinforced sheets for rubber dams intended to be used in water bodies like river should be multiplied imparting strength to meet requirements.

Basic assembly of a reinforced rubber sheet is depicted in Fig. 4.7: the thickness of each layer, technical rubber formulations for each layer, type and construction of fabric varies with the intended application (i.e., scale of water body and purpose).

During vulcanization, using heat and pressure for build rubber compound imparts shape and enhances performance of final reinforced rubber sheet. Press, autoclave, and continuous are vulcanization methods used to cure reinforced rubber sheet. Steam, electricity, hot air, oil are sources of heat and pressure during vulcanization process. Steam is

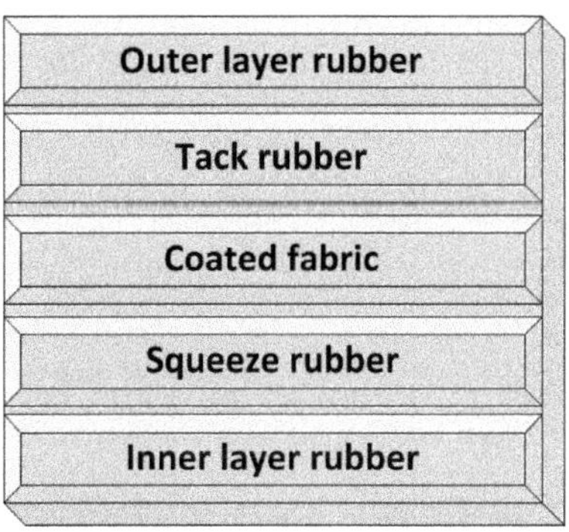

Figure 4.7 Typical assembly of reinforced rubber sheet for rubber dam.

most economical sources for heat and pressure, used in autoclaves and continuous vulcanization (Bhowmick et al., 1994). Both autoclave and continuous vulcanization are popularly used for production of reinforced rubber sheet. Vulcanization parameters—time, temperature, and pressure—are determined well before using a cure meter.

Limitation on sizes of reinforced rubber sheet is resolved using techniques of joint; length of rubber sheets can be increased by making use of mechanical fasteners or vulcanized splicing.

Results obtained from quality test on laboratory scale are important parameters determining the performance of the reinforced rubber sheet in field. Quality test as mentioned in Fig. 4.8 should be performed. Defects mentioned in Figs. 4.9 and 4.10 should be checked during final inspection of reinforced rubber sheet. After performing inspection, the reinforced rubber sheets are packed as per standards or as agreed with customer.

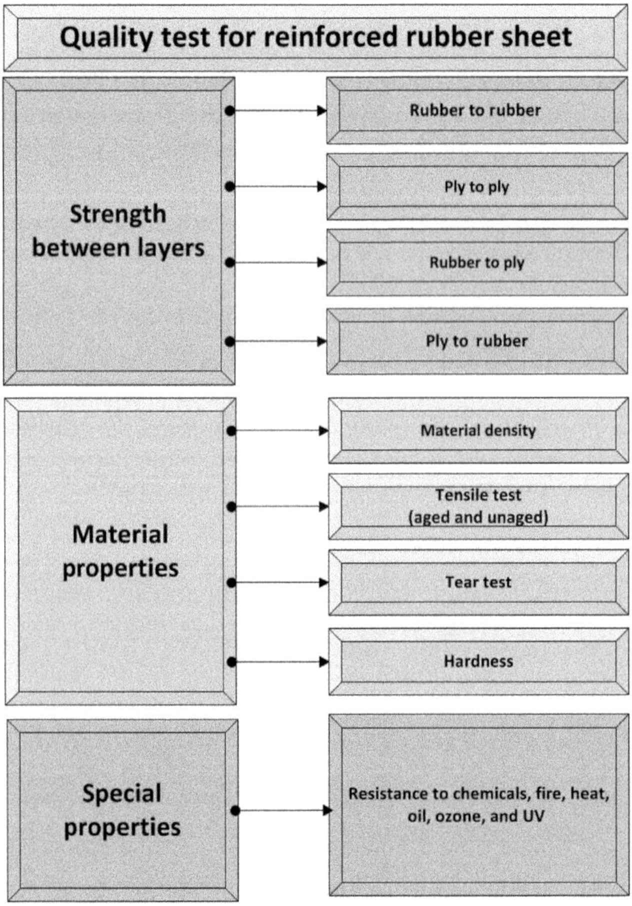

Figure 4.8 Quality test for reinforced rubber sheet for rubber dam.

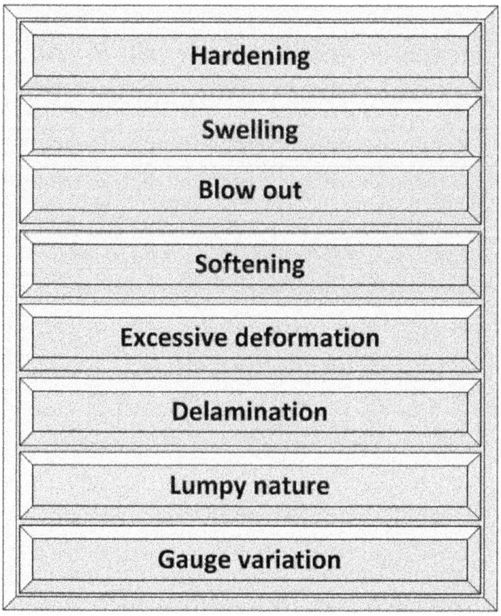

Figure 4.9 Bulk defects to be observed in reinforced rubber sheet for rubber dam.

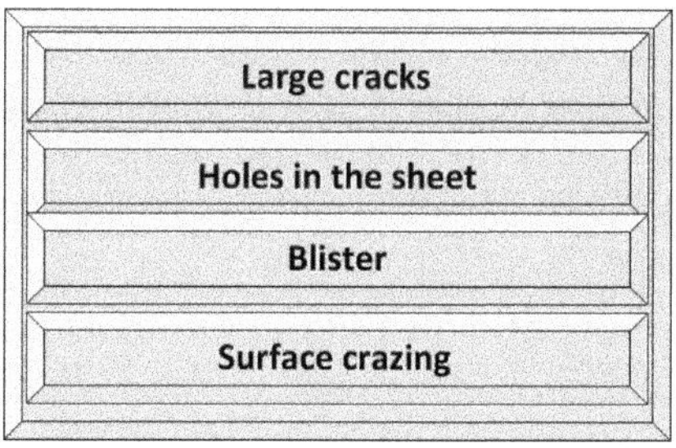

Figure 4.10 Surface defects to be observed in reinforced rubber sheet for rubber dam.

This chapter focuses on manufacturing process for fabrication of individual components (rubber sheet and rubberized fabric) and their assembly. Quality check operations for defects involved in fabrication of a reinforced rubber sheet should be prevented and repaired if present, if not repairable should be rejected from supply lot. As reinforced rubber sheet is major and important component in construction of rubber dam during service, for which specific formulation of the rubber compounds, type of fabric, and specification for processing should be followed to have consistency in the quality of reinforced rubber sheet. Fixation of reinforced rubber sheet in field (i.e., construction technology for rubber dam) includes concrete foundation, control room for mechanical and electrical equipment, and inlet and outlet piping system that is an important part of entire rubber dam during service. Preparatory phase for construction of a rubber dam involves civil engineering and customization of a rubber dam sheet; civil engineering involves inspection of land topography and its suitability for rubber dam, construction of reinforced concrete foundation and piping system. Customization means to confirmation if the rubber bladder is fabricated as per design of the bladder manufacturer to have superior performance, efficiency, and reliability of rubber sheets during service (http://www.rubberdam.org/technology/construction-inflatable-rubber-dam.html). Installation process involves cleaning of rubber bladder and base plate, placing the gasket, installing water tank, laying the rubber bladder on concrete foundation, lifting and expanding the rubber bladder, making arrangements for system of pipes, ditches, canals, tunnels, and supporting structures used to convey water from its source to its main distribution point, positioning the rubber bladder, then inspecting the rubber dam during service. Before installation of rubber bladder on concrete foundation, measuring the concrete strength of the foundation, to see whether it meets the design requirements, is important, secondly the surface of concrete foundation should not be left rough, it is best to keep it smooth. Checking on the pipeline of filing and drainage for "no leakage" for smooth functioning of rubber dam is must. Markings on the base plate, such as center line and dam axis, should be done to avoid offset while fixing rubber dam on the concrete foundation (http://www.rubberdam.org/technology/construction-inflatable-rubber-dam.html). Detailed information on installation of rubber dam is available in reference (http://www.rubberdam.org/technology/construction-inflatable-rubber-dam.html) and Chapter 5, Bonding and Anchoring System for Rubber Dam, and Chapter 6, Inflation and Deflation of Rubber Dam, of this book.

References

Bhowmick, A.K., Hall, M.M., Benarey, H.A., 1994. Rubber Products Manufacturing Technology. CRC Press.

David B.W., 2001. The Application of Textiles in Rubber, Rapra Technology Limited, United Kingdom, ISBN: 1-85957-277-4.

Erman, B., Mark, J.E., Roland, C.M., 2013. The Science and Technology of Rubber, fourth ed. Elsevier, Amsterdam.

5 Bonding and Anchoring System for Rubber Dam

Joygopal Jena
Department of Civil Engineering, Gandhi Institute of Technological Advancement (GITA), Bhubaneswar, India

Abstract

The technology of using rubber as prime construction material for water retaining structure has emerged in very near past. To substitute conventional water storage or diversion structures, inflatable rubber dams are being used as water control structures for more than 60 years. The concept of rubber dam was originated by N.M. Imbertson of the Los Angeles, Department of Water and Power in the 1950s and manufactured by the Firestone Tire and Rubber Co. as Fabridams. The world's first inflatable rubber dam (Fabridam) was installed on the Los Angeles River, California, in the year 1957 for groundwater recharge and flood mitigation, while the Japanese installed their first one in 1964 (Guidelines for Evaluation of Hydropower projects, Federal Energy Regulatory Commission, Ch-10.6). Initially, rubber dam was filled with water, and the inflation was regulated to maintain a desired height by a siphon. Air-filled rubber dam was first introduced by Bridgestone Corporation in 1978. Afterward, many improvements have been made in the construction aspects and thousands of such dams have been constructed and currently in operation in various countries. The Japanese follow their standards of "Land Development Technical Standard of Rubberized Fabric Inflatable Weirs, August 1983 (LDTRC, 1983)." There is no Indian standard for such types of dam. Rubber dams are very light in weight when compared with other controlling/diversion structures of similar height, relatively inexpensive, versatility in application as structures for creating pondage, and/or regulating flow of water in a particular stream or waterway.

Keywords: Rubber dam; dam body; water-filled dam; air-filled dam; anchoring; clamping

Hydraulic Rubber Dam. DOI: https://doi.org/10.1016/B978-0-12-812210-5.00005-5

5.1 Introduction

The technology of using rubber as prime construction material for water retaining structure has emerged in very near past. To substitute conventional water storage or diversion structures, inflatable rubber dams are being used as water control structures for more than 60 years. The concept of rubber dam was originated by N.M. Imbertson of the Los Angeles, Department of Water and Power in the 1950s and manufactured by the Firestone Tire and Rubber Co. as Fabridams (Chanson, 1996). The world's first inflatable rubber dam (Fabridam) was installed on the Los Angeles River, California, in the year 1957 for groundwater recharge and flood mitigation, while the Japanese installed their first one in 1964 (Guidelines for Evaluation of Hydropower projects, Federal Energy Regulatory Commission, Ch-10.6). Initially, rubber dam was filled with water, and the inflation was regulated to maintain a desired height by a siphon. Air-filled rubber dam was first introduced by Bridgestone Corporation in 1978 (Chanson, 1996). Afterward, many improvements have been made in the construction aspects and thousands of such dams have been constructed and currently in operation in various countries. Japanese follow their standards of "Land Development Technical Standard of Rubberized Fabric Inflatable Weirs, August 1983 (LDTRC, 1983)." There is no Indian standard for such types of dam. Rubber dams are very light in weight when compared with other controlling/diversion structures of similar height, relatively inexpensive, versatility in application as structures for creating pondage, and/or regulating flow of water in a particular stream or waterway.

A rubber dam is a permanent water retaining structure made of a sheet of rubber-coated fabric (rubber body). This is typically manufactured from a flat sheet of rubberized fabric by folding it into a tubular shape to be filled with water or air for retaining a designed shape and installed on a reinforced concrete foundation/block using suitable anchoring arrangements. An inflatable rubber dam body is normally manufactured in the workshop/factory as per the design and requirement of shape and size. Manufacturing the dam body in the factory has got its own advantages. A factory contains all infrastructures for handling the equipment and the materials for production and assembly. Entire length of the dam body is divided into parts in accordance with the suitability of the dam site and convenience of erection of the same. The dam body made of rubber is considerably lighter and after manufacturing and testing, the sheets are transported to the site for erection. A concrete foundation is laid on the river/stream bed at a location suitable from hydrological and geological consideration for fixing the dam body. The dam body is fixed on a horizontally laid reinforced concrete foundation block/slab designed to resist the probable static and dynamic loads. The rubber sheets are secured with structural steel members and anchor bolts for its stability. The system of anchorage varies as per the size, depth of water retention, and category of the dams.

5.2 Category of Rubber Dams

Rubber dams are categorized from various considerations as described very briefly as follows. Each type of rubber dam has its own advantages and disadvantages.

5.2.1 According to Medium of Fill

Rubber dams are divided into two categories as per medium of filling material.

1. *Air rubber dam:* where air is used to inflate the dam body. Obviously, this category of dams is lighter, and load on the foundation is lesser, thus foundation cost becomes lesser. But due to its light weight, it is subjected to vibration due to hydrodynamic forces.

2. *Water-filled rubber dam:* in which rubber dams are inflated by pumping water into the rubber body until the design height or pressure is achieved. These are comparatively heavier and more stable with less vibration during its operation.

Typical arrangements of both categories are shown, respectively, in Figs. 5.1 and 5.2. These can be deflated by releasing the water or air filled inside the rubber body by provision of a suitable permanent arrangement. Thus, flood can be allowed to pass over the deflated dam when needed.

5.2.2 According to the Side Slope

The rubber dams are also divided into two types according to the side slope: (1) pillow-like dam in case of vertical side slope and (2) inclined dam, in case of variable side slope.

1. *Pillow-like rubber dam*: This type of dam is provided in case of vertical side slope. However, the vertical side slopes have certain disadvantages: (1) distraction of the normal flow of water when released

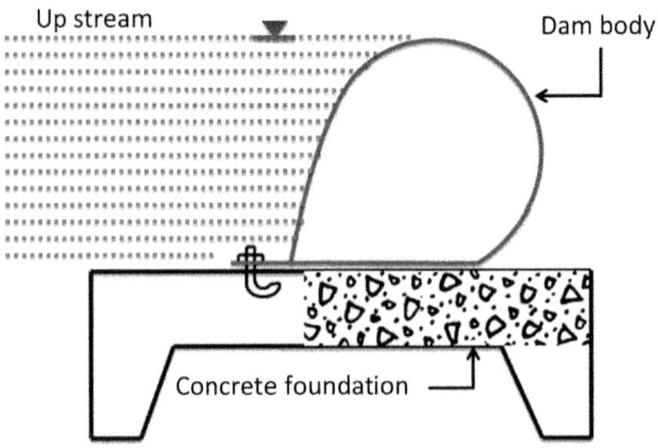

Figure 5.1 Air-filled rubber dam.

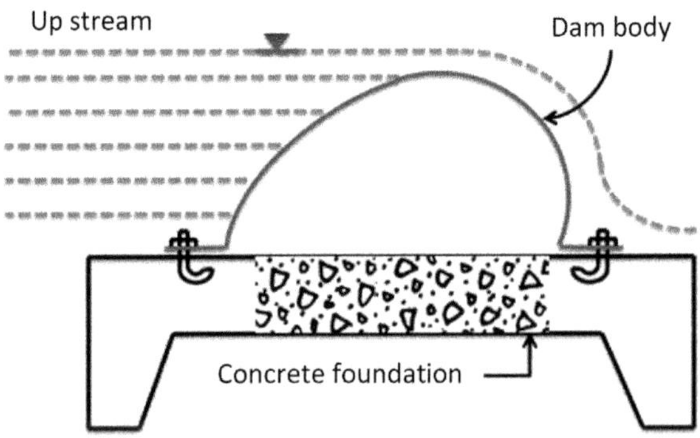

Figure 5.2 Water-filled rubber dam.

causes fluctuation in the downstream flow condition affecting river bed state, especially during inundation, (2) vertical side walls require more structural members for support during construction, relatively needing more reinforcement and hence face more difficulty in construction than sloped side walls, (3) abutment or side walls with vertical slopes also do not become compatible to the nature and disturb esthetics of stream/river, and (4) it is noticed that rubber dams with side vertical walls are more susceptible to damage due to obstruction of upstream flow.

The anchoring of the dam to the foundation also depends on this categorization. In pillow-like dam, the rubber body is not fixed to the side walls of the piers or abutments. This is shown in Fig. 5.3.

2. *Inclined rubber dam*: Inclined dams are provided in case of variable side slope. Rubber dams can be installed in streams/rivers, with any side slope angle for which modification is not necessary to the river

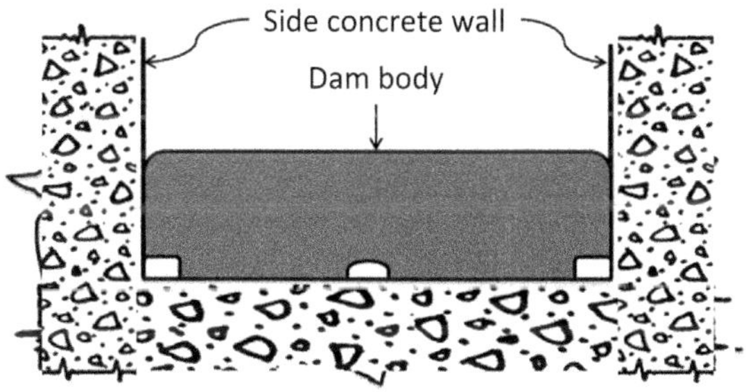

Figure 5.3 Pillow-like rubber dam.

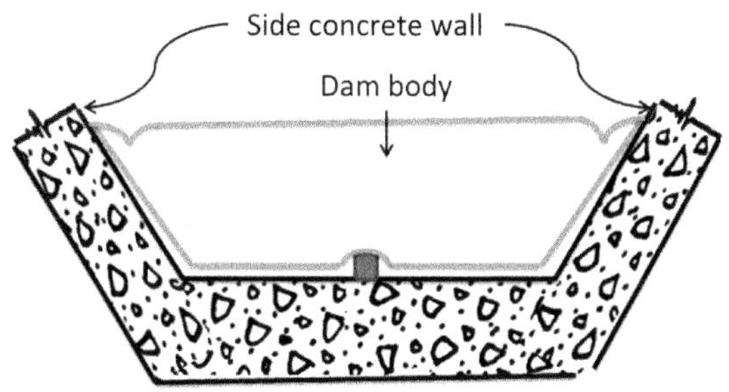

Figure 5.4 Inclined rubber dam.

banks, while in case of steel gates, it is mandatory to provide vertical side slopes for their installation.

In case of such type, the rubber bladder matches the cross-section of the river or the waterway of canal and it is fixed to the bottom concrete foundation and also to the side walls by anchors as shown in Fig. 5.4.

5.2.3 According to the Method of Anchoring

Another way of classifying the rubber dams is on the basis of the method of anchoring.

5.2.3.1 Single Anchored Rubber Dam

In this type, an inflatable rubber dam body is installed on a concrete foundation by fixing it with the help of plates and anchors positioned in a single row as shown in Fig. 5.5. Single anchored dams are of very low height, where water retention is limited to 2 m.

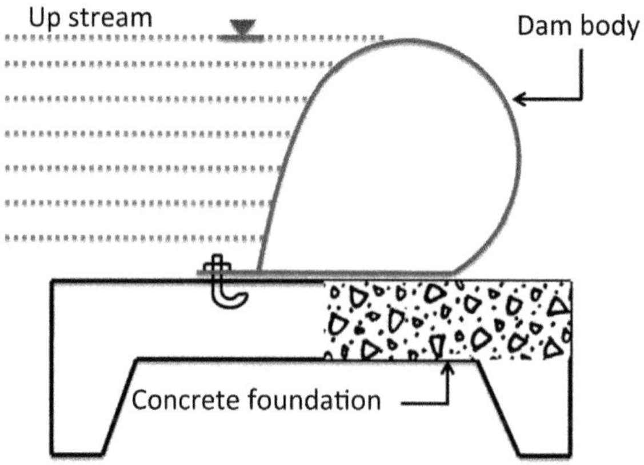

Figure 5.5 Single anchored rubber dam.

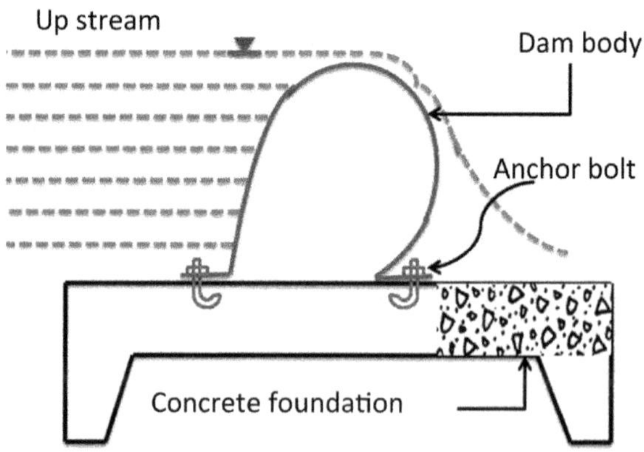

Figure 5.6 Double anchored rubber dam.

5.2.3.2 Double Anchored Rubber Dam

In this case, the dam body is fixed to the foundation in two rows of anchors as shown in Fig. 5.6. Double anchored dams generally are of more than 2-m height.

For the dam with two rows of anchors, base length of the dam is well defined. The distance between the anchor lines should be provided in such a way that the membrane of the downstream face does not lie flat on the foundation face.

5.2.4 Classification Based on the Seam of Rubber Bag

Rubber dams are classified into another two types, depending on the seam of rubber bag.

5.2.4.1 Seamless Rubber Dam

Seamless rubber dam is manufactured in the factory without any transverse joints, that is, with a single piece rubber bag between the sides. This is mainly used on slope dams of over 4 m high. When water is filled to maintain the height of over 4 m, the dam body will not only be subjected to lateral strain on its slope but also high longitudinal strain and shear force. Seamless rubber dam is formed by applying heat pressure in the factory so that the problem of parting of seam joints does not occur.

5.2.4.2 Seamed Rubber Dam

Seamed rubber dam is formed by adhesion of pieces of adhesive rubber after grinding. Due to human factors and the limitation of technology, the seam joining part may retain defects and gets easily damaged under complicated loading condition in the slope section. This will cause the tearing of the dam bag under serious condition and will affect the life of dam bag.

5.3 Bonding of Dam Body

The body of a rubber dam is a rubberized fabric membrane, which is composed of numerous layers of synthetic rubber along with the layers of synthetic fiber reinforcement bonded together firmly in a sandwich construction by vulcanization process in order to provide the required tensile strength. Chloroprene rubber (CR) is employed for its known resistance to climate conditions, ozone, wear and tear, etc. A layer of stainless steel mesh or ceramic chips can be fixed on the surface layer for preventing damage due to vandalism (Chanson, 1996). The layers of rubber-coated fabric are joined together in the longitudinal direction. Height of the dam and the required designed strength dictate actual number of layer of rubber-coated fabric to be provided for a rubber dam body.

During manufacturing of a rubber bladder, quality control becomes very vital which is to be maintained throughout the processes of rubberizing, joining together, and pasting. Unevenness or local imperfections on the above or discontinuity in gluing may lead to separation of layers of bladder. In some cases, where the bladder is too large, transportation becomes a problem imposing restrictions on packing dimensions. This necessitates the adhesion of the bladder to be carried out on site. In such case, special attention should be given to quality management, since the dam site environments (e.g., dust and humidity) may adversely affect the adhesion process.

5.3.1 Specification of Rubber Dam

The quality of rubber used for manufacturing the rubber dam is very important. There are no fixed specifications for the rubber dam as these are tailor made to suit the actual site condition. The specification of a typical type of dam is decided as per design consideration adopted for such rubber dam. Strength being the major criteria for the dam, general specification with respect to allowable stress of the dam body is described in this section.

The circumference of a water-filled dam is about 4.8 times the maximum height while in case of an air-filled dam, it is 3.5 times. To accommodate the dam body, the foundation of water-filled dam is wider than air-filled dam of same height (Zhang, 2007). A minimum thickness of 10 mm is to be provided for the rubber dam body and it is to be reinforced by at least two layers of nylon canvas. Tensile strength in the direction water flow after heat aging at 100°C for 4 days shall be more than 80% of initial strength and safety factor shall be more than 8 under normal loading condition. In longitudinal direction, it should be more than two-thirds of that in the direction of flow. Adhesion between nylon canvas and rubber shall be more than 6 and 4 kg/cm initially, under hot water resistance test at 70°C for 4 days, respectively. Galvanized steel or stainless steel should be used for the clamping materials and anchoring bolts and shall have at least two ridges to form dual sealing line to prevent any leakages. Double anchoring line is essential to accommodate to the situation when upstream water level is higher than downstream water level or vice versa. Safety factor of clamping bolts should be more than three times against tensile strength of materials used under normal loading condition.

5.3.2 Physical Property of Rubber Dam

Ethylene propylene diene monomer (EPDM) compounded rubber with nylon reinforcement or CR or equivalent with nylon-reinforcing fabric is recommended for use for the dam body. External surface of the dam body should be abrasion resistant. The external rubber cover thickness shall not be less than 4 mm which forms part of the rubber body (total thickness shall not less than 10 mm) and the surface must be textured to reduce the possibility of the debris dragging across the surface of the rubber body during the fully deflated stage. It should gain high water resistance for tropical area, high durability and capability of operating in accordance with the specified conditions, abrasion resistance, tear resistance, impact resistance, and ultraviolet resistance. When adopted, fin or deflectors shall be formed as an internal part of the rubber body during manufacturing of the rubber body, lamination to the rubber body after manufacturing should not be allowed. The clamping plates and anchor plates should be designed to fix uniformly the body and transmit hoop tension of the body to the concrete foundations. The clamping plates and anchor plates should have complete water and airtight construction at both inflated and deflated conditions of the body. It should be liable for deterioration of the rubber materials throughout the life of the rubber dam.

5.3.3 Considerations for Construction of a Rubber Dam Body

Detailed design of the rubber dam body is not discussed in this chapter. The following forces are considered while designing a dam body. These are (1) upstream hydrostatic force, (2) internal water/air pressure, (3) weight of rubber membrane, (4) downstream hydrostatic force, and (5) hydrodynamic forces. El-Jumaily and Salih (2005) have analyzed the stresses in the membrane with only hydrostatic condition. The following aspects should be considered during construction of a rubber dam. The material should be sufficiently strong enough to resist tearing/puncture and the quality should not deteriorate when exposed to nature. Normally, the body of the dam is manufactured as a single unit to fit to the span of the dam to avoid the requirement for site welding of seams. Rubber dam bladders are made of high strength canvas and rubber layer to make it air/watertight. Rubber sheets include three layers: outer layer, middle layer, and inner layer. Outer rubber layer should have heat-resisting, wear and tear-resisting, and ozone-resisting characteristic. Middle layer takes the tension of the canvas/reinforcing layer and joins these layers. Inner layer serves the main function of making the dam body watertight and airtight to retain its shape and size.

5.3.4 Durability of the Rubber Dam

The required strength can be achieved by using materials of varying thickness. However, practice has shown that using thicker cover has certain advantages with respect to durability. Thinner cover is particularly susceptible to puncture and tearing damage from hydraulically transported materials. Durability is also improved by use of best quality of rubber sheet and strong nylon fabric. Oxygen and ozone attack elastomers and alter their microstructures by causing chain scission and cross-linking. In addition to this, increase in temperature, ultraviolet rays, development of stress, and presence of some metals (e.g., copper, cobalt, and manganese) accelerates oxidation. Synthetic materials, e.g., CR and EPDM, have been used to enhance dam materials' resistance to weathering and ozone. The EPDM can withstand temperature variations from -100 to 180°C.

5.4 Anchoring System

As discussed above, depending on the type of dam, the "pillow-like dam" and the "inclined dam," the rubber dam is fixed to the side walls or not. The anchorage system also varies accordingly as discussed briefly here.

5.4.1 Anchorage of Dam Bag With Concrete Base

Various types of anchorage systems are adopted around the globe. Types of the anchoring vary whether the dam is an air-filled or a water-filled one. The water-filled dams being heavier are more stable. The air-filled dams are susceptible to more vibrations. Therefore, anchoring of the inflated rubber dams is very important.

5.4.1.1 Single-Line Anchoring

In case of a stream with very low discharge and minimal tail water conditions, where dam height is less than 2 m, the base length of a rubber dam body is fixed to the foundation by a single row of anchors along the longitudinal direction of the dam, that is, the transverse length of the membrane, which lies flat on the foundation surface. The base length is decided as per varying load conditions.

5.4.1.2 Double-Line Anchoring

Where the height of the inflatable dam is more, that is, more than 2 m, the body of the dam should be fixed on the foundation with two rows of anchors to facilitate reduction of load on individual anchors. The double-line anchoring system (Fig. 5.6) can increase the stability and reduce the vibration of the dam body. At dam sites where reverse flow occurs, it is always advised to adopt double-line anchors.

5.4.2 Anchorage of Rubber Dam Sheets With Side Walls of Abutment or Piers

As discussed earlier, while in case of a pillow-like rubber dam, the dam bag is not fixed sideways, the free end of an inclined dam is fixed to the side walls of the abutment or an intermediate pier after sealing the membrane in shape of an arc or a straight line. In some cases, an inflatable dam body is extended beyond river bank/water way where the ends are fixed with separate special arrangement. Different manufacturers follow their own method of anchoring system. Type of the anchoring also depends on the height of the dam or water retention. There are numerous types of anchors available and they are placed with various anchor angles depending on the type of work and purpose. In case of a rubber dam, anchors are always placed vertically on the horizontal bed of the foundation and fixed with the reinforcement steel prior to the concreting of foundation.

Gurt et al. (2015) have presented design and analysis of inflated dam, where they have proposed the clamping system and defined geometry of the rubber dam at the slope of the weir pier. The appearance of an inflatable rubber dam depends on geometric, hydraulic, and material parameters of the dam. As shown in the schematic diagram below, the geometric parameters are the span "l," the distance between the clamping lines "b," the height "w" of the clamping lines, the small height "h," the angles "β_1" and "β_2," the slope of the weir pier, and circumference "U." The angles β_1 and β_2 are defined as a function of the other geometric parameters as stated.

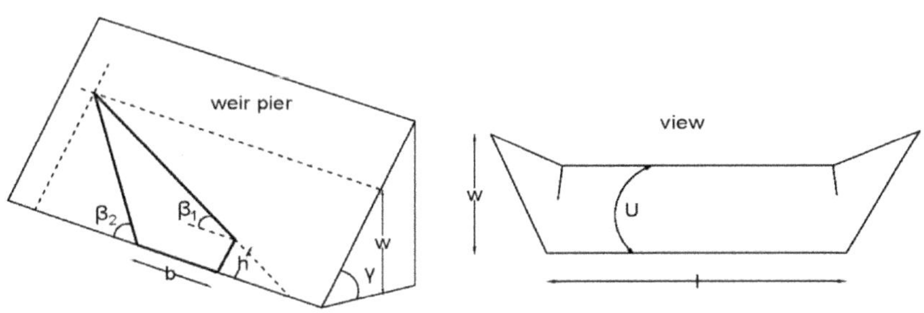

Schematic diagram of clamping geometry (Thick line) on a weir pier.

The height of a dam is determined by the parameter's circumference and the distance between the clamping lines. The slope of the abutment or the side piers is also an important parameter for deciding the overall geometry of the dam.

5.5 Various Types of Anchoring

There is no fixed type of anchoring methods or arrangements. Several manufacturers of the dam devise their own method of fixing the dam to the foundation and walls. These methods are also being developed day by day.

Some of the types of anchoring systems are discussed in the following subsections.

5.5.1 Concrete Wedge Blocks

This is a very old method of fixing or anchoring of rubber sheet to the concrete foundation block or slabs. This system was adopted initially in China (QHEPIL, 1999). In this system, slots are provided in the foundation base slab. The ends of the rubber sheet are folded and inserted into slots with the hand. A wooden block of circular section is inserted into the fold of the rubber strip which helps to keep the rubber strip in position and serves as a lock. The sheet is anchored by two sets of precast concrete wedge blocks such that the front wedge keeps the rubber sheet in position butting against the upstream face of the concrete slot/groove and the back-wedge block remains in position by hammering into the block. The slots and the blocks should have accurate dimension and sharpness for quality work. This method serves for rubber dams of very low height and for simple hydraulic forces. Wooden anchorage blocks can also be used.

The efficiency of the grip system is poor and there is likelihood of slippages as there is no secondary locking arrangement. This system does not work perfectly and due to its poor results, this method is not very popular.

5.5.2 Steel Clamps With Anchor Bolts

The clamping system is an improved method of anchoring of the rubber dam body to the foundation block. It also helps sealing the bottom of face of the dam body and makes it air and watertight. This kind of system makes the installation of the dam much faster, simpler, and more efficient. As discussed earlier, the rubber dam bag may be fixed by a system of single clamping line (upstream side) or a double clamping line (both upstream and downstream sides). In a situation where downstream water level and wave forces are significant, it may be necessary to provide a double clamping line system. The fitting arrangement includes embedment of a plate inside the concrete (lower clamping plate), clamping plate above the concrete surface (upper clamping plate), and anchor bolts.

For fixing the rubber dam body on the concrete foundation/slab, steel clamp elements and stainless steel anchor bolts and nuts are generally used. Fixing of rubber sheet is a difficult and precision work in comparison to other construction works of the rubber dam project. At first, the anchoring slot is kept open for second stage concrete for positioning and fitting and fixing the anchor bolt with dowel bar of the slot.

Lower anchor plate is fixed prior to final layer of foundation concrete keeping the anchor bolt inserted into the holes of the plate and the anchor bolt is fixed in position by welding it with the dowel bars. The level of lower plate is maintained as per the design with the help of leveling instrument and this level is adjusted with the help of bottom nut of anchor bolt. Then, second stage concrete is laid up to the bottom level of lower plate. The thickness of lower and upper plates is designed according to the height of rubber dam. Utmost care is taken during this work for maintaining proper alignment of the anchor bolts, as any deviation of anchor bolts may lead to improper fitting of anchor plate which may create problem for fixing the dam body to the foundation concrete.

In another system of anchoring, the anchor bolts are fixed in position by welding those with the reinforcement of the foundation prior to concreting. After concreting, the bolts are cleaned and fixing blocks are inserted into the bolt. The rubber sheet is punched and then inserted into the extended portion of the fixed bolts. Then, the rubber sheet is tightened with the help of the upper clamping plate. This type of anchoring system is shown in Figs. 5.7–5.12.

5.6 Clamping Rubber Fabric With Concrete

In a general anchoring system, the dam body is fixed to the foundation with clamping plates.

Figure 5.7 Anchor bolt fixed on foundation.

Figure 5.8 Cleaning foundation anchor bolts.

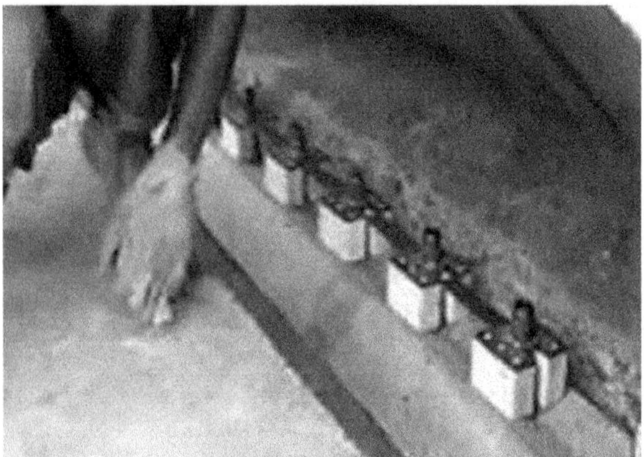

Figure 5.9 Typical fixing blocks.

Figure 5.10 Anchor bolts on walls.

Figure 5.11 Punching rubber sheet for fixing.

Figure 5.12 Clamp plate fixed with rubber sheet.

5.6.1 Clamping by Punching the Fabric

Anchors, where the rubber sheet of the dam body is punched, and gaskets are provided at the top and bottom of the sheets to make it air/watertight. Several methods have been developed in order to secure the fabric to the base slab without punching the hole in the fabric with the help of a special clamping bar and an anchoring system. There are numerous advantages of this clamping system.

5.6.2 Clamping Without Punching the Fabric

By avoiding the punching in the sheet, the stresses in the fabric are minimized by eliminating stress concentration developed due to holes made by punching through the fabric. Eventually, the strain and creep are reduced and the possibility of failure of the sheet by creep is almost eliminated. An airtight seal is also maintained even at higher levels of stress and strain. With this system of anchoring, the risk of failure due to pull out and air leakage is almost negligible.

5.6.3 Clamping Plate and Anchor Bolts

The clamping plate should have very good resistance for rusting. Normally, galvanized steel is used for the purpose. Stainless steel is to be used where saline water is encountered. The clamp and anchor bolt sizes vary with the manufacturer of the rubber dam. The specification of these clamping materials, that is, size and dimensions of clamping plates and anchor bolts, varies primarily with varying height of water retention.

5.7 Foundation for the Rubber Dam

Construction of concrete base slab is very important because all the loads of the dam are transferred to the foundation. This gives a rigid and smooth surface for fixing the rubber sheets for stability of the dam. Preparation surface of the concrete foundation depends on the type of the rubber dam and height of water retention. In some cases, flow of water is reversible. In such cases, the shape of the foundation shall differ. All measuring devices, water/air control systems for inflating and deflating of dam, are generally embedded in the foundation concrete. Details of the foundation design and construction procedures are not dealt in this chapter. However, some salient aspects are discussed very briefly. Accepted criteria for design of the rubber dam have been stated in Section 10-6.4 of Engineering Guide line for evaluation of Hydropower Projects, Federal Energy Regulatory Commission.

5.7.1 Loads Considered for the Design of the Concrete Base

Rubber dam being a permanent structure, strictly speaking, all the forces/loads considered for gravity dam should be considered for design of the concrete base of the rubber dam. But the height of rubber dams (normally, limited to 5 m) is much less than the other standard dams for which the bed width of the dam is also less. The bed width varies from 4.8 times the maximum retention of water. Therefore, some of the major forces such as "wave force", "uplift force," and "earthquake force" may not be considered as in case of other conventional dams, unless required for designated site. Other forces such as (1) hydrostatic force, (2) silt or ice load (whichever applicable), (3) self load of dam in both empty and full condition, (4) hydrodynamic force, and (5) inertia force are considered for foundation concrete design.

5.7.2 Selection of Site for Suitability of Foundation

Dam sites selected after conducting comparative technical and economic investigations according to rubber dam structures and its operation necessitate in consideration of various factors, such as, topography of land in the vicinity, geological features of the site, discharge of water flow, silt contents in water and deposition, and environmental protection. The dam should be built preferably on relatively straight river or channel segment with smooth water flow and stable bank slopes. It is unsuitable to choose the dam site located on the degrading and aggrading river segments. The foundation should be geologically sound. Foundation with hard and compact rock strata is ideally preferred. In case of fractured/soft rock/weak foundation, it should be treated suitably to obtain an impermeable foundation with desired bearing capacity.

5.7.3 General Construction Procedure for Lying of Foundation Concrete

The foundation works are started after making a river diversion arrangement after construction of a cofferdam and making the side channel. For diversion of small discharges, syphon arrangement can be done. This is done for facilitating the excavation and foundation treatment and subsequent lying of concrete. Various tests such as percolation test and test for safe bearing capacity are carried on for finding the suitability of foundation. To reduce seepage loss through the foundation of the dam, various foundation treatment methods are adopted to make it impermeable. In case of fractured rock, cement grouting is done. In permeable foundation (sandy soil), metal sheet piling is done. Concreting is done on the treated foundation as per the design and specification. The concrete should conform to any standard code of practice. In India, BIS specification vide IS 456 latest is followed. As a practice, the downstream of the dam foundation is connected to the river bed in the downstream with a vertical slope of 1:4–1:2 (V:H).

5.8 Special Types of Rubber Dams

5.8.1 Shield-Type Rubber Dam

Shield-type rubber dam is composed of air-filled rubber bladders and steel shield plates which is a state-of-the-art alternative to traditional rubber dam. These are very useful for channels, small spillways, and other waterways with suspended sediment, bed load, and floating debris. This is a modular design and manufactured along with the supporting arrangement which is easy for installation. In this system, each piece of steel support plate corresponds to a rubber supporter. Therefore, during installation, only a specific portion of the dam is positioned and there is no need to bother about the overall position of the inflated rubber dam. So, installation of such dams becomes easier and there is saving of time and labor.

In this system, the repair becomes easier. In case of air-filled dam during repair of any portion of the dam whole body is considered. In shield-type rubber dam, it is easy to repair as only that part of the dam which needs to be repaired is considered, that is, only replacement of the broken part is done instead of considering the whole bladder. The shield-type rubber dam can be installed without central piers. Hence, dam of any span or length can be considered. The hard steel shield plate protects the soft rubber bladder from the sharp objects floating in water.

5.8.1.1 *Steel Shield Plate*

The components of the shield plate are steel panels, ribs, suppression band, etc. Instead of high-strength structural steel, 12 mm thick panel and welding ribs are used. Intercell high-strength composite rubber with soft connection is provided in 3–5 m spacing. These are connected by reinforced rubber webs. The steel shield plate is lowered and raised by the inflatable rubber bladder.

Acknowledgments

The author expresses his deep sense of gratitude and is extremely grateful to the galaxy of authors and institutions whose books and reports/publications have been referred during every stage of writing. Also, the author is extremely thankful to the project authorities of Jhanjhavathi Irrigation project, Andhra Pradesh (where the first rubber dam in India has been installed), for giving an opportunity for physical presence in project site during the installation of the dam.

References

Chanson, H., 1996. Rubber dam hydraulics: some hydraulic aspects during overflow above inflatable flexible membrane dam. in: Report CH47/96, Dept. of Civil Engineering, University of Queensland, Australia.

El-Jumaily, K.K., Salih, A.A., 2005. Analysis if inflatable dams under hydrostatic conditions. J. Eng. Dev. 9 (3), September.

Federal Energy Regulatory Commission, Engineering Guide Line for Evaluation of Hydropower Projects, Chapter 10.6, pp. 10–58.

Gurt, R., Deutscher, M., Gebhrdt, M., 2015. Design and analysis of reinforced rubber membranes for inflatable dams. In: Structural Membranes, Proceedings of VII International Conference on Textile Composites and Inflatable Structures, pp. 306–317.

QHEPIL, 1999. The Rubber Dam and the Technical Standard of Rubber Dam. Qingdao Huahai Environmental Protection Industry Co. Ltd., China.

Zhang, X.Q., 3rd January 2007. "Flexible structures" water power and dam construction, Free Newsletter.

Zhang, X.Q., Tam, P.W.M., Zheng, W., 2002. Construction, operation, and maintenance of rubber dams. Can. J. Civil Eng. 29 (3), 409–420.

Further Reading

Anwar, H.O., 1967, Inflatable dam. J. Hydraulic Division, ASCE, 93 (HY3).

Azizul Hoque, S.M., 2010. Performance Evaluation of Pekua Rubber Dam at Cox's Bazar. Department of Water Resources Engineering Bangladesh University of Engineering and Technology, Dhaka, Bangladesh, Thesis).

Chanson, H., 1998a. Hydraulics of rubber dam overflow: a simple design approach. In: THOMPSON, M.C., HOURIGAN, K. (Eds.), Proc. 13th Australasian Fluid Mech. Conf., Melbourne, Australia, 13–18 December, Vol. 1, pp. 255–258.

Chanson, H., 1998b. Use of rubber dam in flood mitigation of Hong Kong. J. Irrig. Drain. Eng. May/June/1998.

Jena, J., 2012. Rubber dam and its prospects. In: Proceedings of International Conference on Structure and Civil Engineering, Bangalore, ACEE, DOI: 02. SCE 2012.02.11, November 2012.

Mason, D. The Inflatable Rubber Dam in Water Resources Management. Sumitomo Electric Europe Ltd, 30 Dorset Square, London NW1 6QJ, UK.

Patil, A., 2016. Inflation and Instabilities of Hyper Elastic Membranes (thesis). Dept. of Mechanics, Royal Institute of Technology SE-100 44 Stockholm, Sweden.

Sumitomo Electric Industries Ltd., 1997. Pioneering Rubber Gate Technology. Sumigate, Osaka, Japan.

Shivhare, V.K., Bansal, A., Verma, K., 2016. Rubber dam: design and construction aspects under Indian environment. Int. J. Eng. Technol. Sci. Res. 3 (5), ISSN 2394–3386.

Parish, Y., 2016. Evaluating the design, construction and use of rubber dams. Int. J. Adv. Biotechnol. Res. (IJBR) 7 (1), 159–166. ISSN 0976-2612, Online ISSN 2278–599X.

6 Inflation and Deflation of Rubber Dam

***Arun Kumar*[1] *and Shahid ul Islam*[2]**

[1]Alternate Hydro Energy Centre, Indian Institute of Technology, Roorkee, India,
[2]Baba Ghulam Shah Badshah University, Rajouri, India

Abstract

Inflatable weirs, also known as rubber dams, are flexible elliptical structures made of rubberized material attached to a rigid concrete base and inflated by air, water, or a combination of air and water. Those are made of durable fabric and are pressurized by air, water, or a combination of both. After rubber dam is inflated, it serves as weir. Deflation of rubber dam during floods helps to flush out sediments automatically and prevents the damage due to flood. Rubber dams are simple, reliable, and flexible structure resulting in wide range of application in hydropower, irrigation, and flood control and water resource management. The objective of this chapter is to provide the details of inflation and deflation, hydraulics and structural design of rubber dam. The details of installation of rubber dam are also given. Optimization of suitable inflation/deflation mechanism as well as and details of safe and user-friendly operation are also discussed in this chapter. Problems associated with inflation and deflation and their mitigation is also provided.

Keywords: Inflatable dams; Rubber dam; SHP; Weir; Inflation; deflation

6.1 Introduction

Inflatable weirs, also known as rubber dams, are flexible elliptical structures made of rubberized material attached to a rigid concrete base and inflated by air, water, or a combination of air and water. Inflatable flexible membrane dams (IFMDs) are long tubular-shaped fabrics placed across the watercourses to provide the temporary storage, so that water can be diverted and regulated in the off taking canal and are also called rubber dams or fabridams. They are made of durable fabric and are pressurized by

Hydraulic Rubber Dam. DOI: https://doi.org/10.1016/B978-0-12-812210-5.00006-7

air, water, or a combination of both. After rubber dam is inflated, it serves as weir. Deflation of rubber dam during floods helps to flush out sediments automatically and prevents the damage due to flood. Rubber dams are simple, reliable, and flexible structure resulting in wide range of application in hydropower, irrigation, and flood control and water resource management (Sahoo et al., 2010a) (Fig. 6.1). The objective of this chapter is to provide the details of inflation and deflation of rubber dam. The details of installation of rubber dam are also given. Optimization of suitable inflation/deflation mechanism as well as and details of safe and user-friendly operation are also discussed in this chapter. Problems associated with inflation and deflation and their mitigation is also provided.

The first rubber dam was developed by an American engineer, Norman Imbertson, during mid-1950s. The height and length of which was 1.52 and 39.6 m, respectively. Initially it was called as "Fabridam," and inflation was done by both water and air. It was made of nylon-reinforced neoprene and was manufactured by Firestone Tire and Rubber Co., Akron, Ohio (Imbertson, 1960). Rubber dams are of long spans and are adaptable to different side slopes. Apart from low life cycle cost of rubber dam, the construction time is short, and maintenance and repair is easy. Moreover, they are adaptable to adverse conditions and are earthquake resistant (Zhang, 2007). The simplicity, reliability, and flexibility of the inflatable dam structure result in its wide range of applications (Sahoo et al., 2010a).

The selection of type of rubber dam is site specific. It is not always possible to select most desirable rubber dam for a particular system. The rubber dam is designed to suit to the local conditions, which include topography, geology, ecology, social concerns, and availability

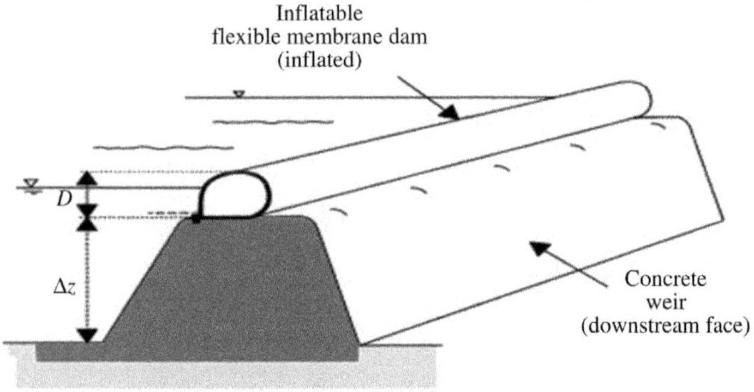

Figure 6.1 Sketch of a rubber dam (Sahoo et al., 2010a).

of material. Rubber dam changes its shape with increasing inflation. Rubber dam is selected based on the design parameters for the particular site. Design parameters include membrane length, anchorage distance, and inflation pressure (Watson, 1985). There is a relationship between the governing parameters and shape of inflatable weir (Hitch and Narayanan, 1983). When the water in the stream is more, rubber dam is deflated (Rantz et al., 1982). Inflatable dam is a feasible alternative to return river to its natural condition which is recent trend (U.S. Army Corps of Engineers, 2001). Inflated weir is more economical than other conventional weirs (Hasan and Kabir, 2014; Ul Islam and Kumar, 2015, 2016). Sarker et al. (2011) discussed the performance evaluation of the inflatable weir project for irrigation development. The problems associated with rubber dam construction were reviewed by Tam and Zhang (1999).

The problems associated with inflated weir include vibrations, vandalism, and floating debris (Khatsuria, 2004). Air-inflated dams are easy to inflate, and maintenance is less as compared to water-inflated dams. In air-inflated dams, vibrations take place when water passes above the light weight dam. These vibrations result in premature dam failure because of the abrasion effects produced by floating debris that get rubbed with dam body.

The impact of these debris (branch, tree, ice, etc.) cause puncture in the dam body, and this must be avoided at any cost. In case of potential debris impacts, a hard cover (e.g., steel gate) might be required. Inflatable dams are subject to outside damages, particularly vandalism and punctures by fast flowing sharp objects. Damages to the rubber dam body can be classified into four types: (1) small puncture, (2) surface damage, (3) small area damage, and (4) large area damage. Thus, both precautionary and remedial measures should be carried out to keep inflated dams in efficient condition as long as possible (Zhang et al., 2002). The problems like damage due to flow-induced vibration, inflation, deflation, vandalism, and due to objects carried from upstream of the weir were discussed (Tam and Zhang, 1999).

6.2 Parts of Rubber Dam

The main parts of rubber dam are as follows:

1. concrete foundation,
2. flexible membrane,
3. anchoring mechanism,

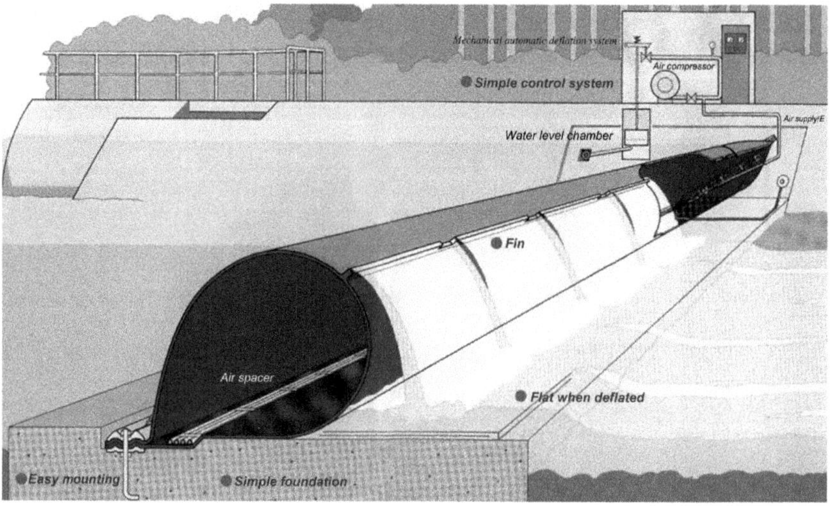

Figure 6.2 Cross-section of an inflatable dam (U.S. Army Corps of Engineers, 2001).

4. pipe system for inflation and deflation, and
5. a pump for inflation of water-filled inflatable dam or air compressor in case of air-inflated rubber dam.

Rubber body is protected to a concrete foundation by its anchor line as shown in Fig. 6.2.

Besides the four main components, there are some other components, which are mentioned below:

1. upstream and downstream concrete floor and protection works (apron),
2. massive concrete base (dam base),
3. filter material,
4. slope wall and protection works,
5. inverted filter at river bed, and
6. bank protection work.

6.3 Medium of Inflation

Rubber dam can be inflated by air, water, or combination of both. Depending on the inflation medium, they are classified as following:

1. air-inflated dam,
2. water-inflated dam, and
3. combined air- and water-inflated dam.

Suitable medium of inflation for rubber dam is selected for a particular site considering the following:

1. site conditions (ambient temperature, quality of foundation, and quality of fill water availability),
2. purpose of rubber dam and operation policy (required function, operation frequency),
3. control and maintenance, and
4. economics.

Option of water for inflating and deflating is subject to the availability of good-quality water in the required quantity. Air media–based dams are operated at faster speed and require lesser power.

6.4 Advantages of Rubber Dam

The main important advantage of rubber dam is that longer span can be made as compared to conventional weirs. Also, when deflated, they can be laid flat on river bed without hindering the flow of water in stream. The ease and flexibility of the inflated dams and its demonstrated reliability are the main reasons for its wide-scale applicability. The inflatable dams are cost-effective and have minimal moving parts. They are well adapted to freezing water condition and are easy to operate for maintaining upstream water level without breakup of ice cover and hence prevent the downstream side from ice jam flooding (U.S. Army Corps of Engineers, 2001). The inflatable weir has proven reliability, and its flexible structure helps to control flood easily (Sahoo et al., 2010b).

The advantages of rubber dam or inflatable rubber weir over conventional weir in terms of ability to inflate or deflate, lighter foundation, longer span without piers, sedimentation resistance, faster installation, and minimum construction work make the demand for inflatable weir increasing. Following are advantages of rubber dams (Zhang, 2007):

1. Long span and adjustable to diverse side slopes: These can be installed in wide rivers and are feasible for any slope angle.
2. Short construction period: Compared with conventional weirs, construction of the inflatable dam is faster. The dam material is lighter and can be handled easily. Moreover, it

can be fabricated and rolled in single piece at a factory for convenient transportation to inflatable dam site.

3. Easy maintenance and repair: Minimal maintenance is needed for inflatable weir. Painting, greasing, or lubrication is not required. Any damage to the weir is early repairable.
4. Low project life cycle cost: The life cycle cost of an inflated weir because dam body is low due to prefabricated components, very less modification to river banks, operation is easy, less maintenance, and faster construction.
5. Earthquake resistant: The rubber membrane is of less weight as compared to conventional weir due to which very less foundation work need to be done which result in earthquake-resistant structure.
6. Environment friendly: The crest-adjustable inflatable weir forms a beautiful impoundment in otherwise fluctuating waters. This creates a stable aquatic habitat and provides recreational opportunities for various water activities.
7. Adaptable to adverse conditions such as cold climatic condition and floods: The rubber dam is operable in very cold climatic conditions where a steel gate is difficult to operate. Furthermore, a rubber dam also reduces the problem of flooding upstream due to the buildup of ice at a conventional hydraulic structure.

6.5 Installation Method

Rubber dam has the advantage of the simplicity and less installation time. No matter what the size of the inflated dam, it is delivered to site in one piece and ready for installation. It is simply unrolled and the anchored to the foundation. No need of heavy lifting equipment or specialized plant. This makes it particularly suitable for remote sites or developing countries. The whole operation can be completed in a matter of days depending on the dam size. For installation of rubber dam, the angle of head wall should be 127 degree and angle of anchorage will be 17 degree (Yooil Engineering Co., Ltd). There should be inlet for water to be filled inside rubber dam in case of inflation and outlet for discharging water from the rubber dam in case of deflation. Specific skill is required for installation of piping work. The nuts and bolts should be of antirust and water tight material. The angle plate must be of waterproof

Figure 6.3 Sequential steps for installation method of rubber dam (Yooil Engineering Co., Ltd).

for it always remains under water (Yooil Engineering Co., Ltd). The inflatable dam is mainly suited to sites wherever the width of the structure is large as compared to its height. Single spans over 100 m in width are achievable with only small size intermediary concrete piers. Installation process as is shown in Fig. 6.3 shows initially concrete foundation body is constructed, and thereafter anchor, clamping plates, and pipes are installed. After that on the top of the concrete foundation, rubber body is installed. Finally control room and supervisory control and data acquisition (SCADA) system is installed. Compared to a conventional weir, the inflatable weir is light in weight and handling is easy. The construction of the concrete foundation of inflatable dam is lighter because of lesser load. In alluvial river, the foundation is to have deeper foundation, similar to conventional weir up to after scour depth. Installation of the inflatable weir is quick, easy, and economical process. The step-by-step method for installation is as follows:

1. foundation work,
2. installing embedded plates and anchor bolts,

3. installing embedded nozzles, pipes, and drain pipes,
4. casting of concrete,
5. lifting rubber body,
6. unrolling rubber body,
7. fixing anchor bolts with rubber body,
8. installing the control equipment,
9. inflation and deflation test, and
10. commissioning.

6.6 Inflation and Deflation Phenomenon of Rubber Dam

Inflation medium of rubber body can be compressed air or pressurized water. In the control room, the devices to inflate and deflate the rubber dam are installed. The inflation of rubber dam is done with air blower/compressors in case of air-filled rubber dam, while in case of water-inflated dam, inflation is done using hydraulic pump. Other ancillary devices such as valves are also used for inflation and deflation process.

There are three types of deflation systems: bucket, float, and electrical. In bucket type, when the water level rises to the deflation level, water flows inside the bucket resulting in its upstream fall due to which exhaust/exit valve opens resulting in deflation of rubber dam. In case of float type, a float fluctuates up and down with the rise and fall of water level, respectively. Exhaust/exit valve automatically opens when water level reaches deflation level. In electrical type, a sensor pressure or ultrasonic or radar type of water level is used to monitor the level of water. When the water level reaches deflation mark, exhaust/exit valve automatically opens (Sahoo et al., 2010a; Chanson, 1997, 1998). Chanson (1997, 1998) reviewed the instances overflow of IFMDs and provided detailed analysis for both deflated and fully inflated configurations (Chanson, 1997, 1998). A mechanism of cavitating flow separation was also explained. Several forms of fluid-structure instabilities were highlighted, and guidelines were suggested for their mitigation. He also gave an easy design approach for hydraulics of inflated dam overflow. Reinvestigation of overflow characteristics of inflated rubber dam was made. Analytic calculations were developed to predict the nappe trajectory above inflated dam, as well as the free-falling jet, and results were compared with over 160 experiments. The results provided

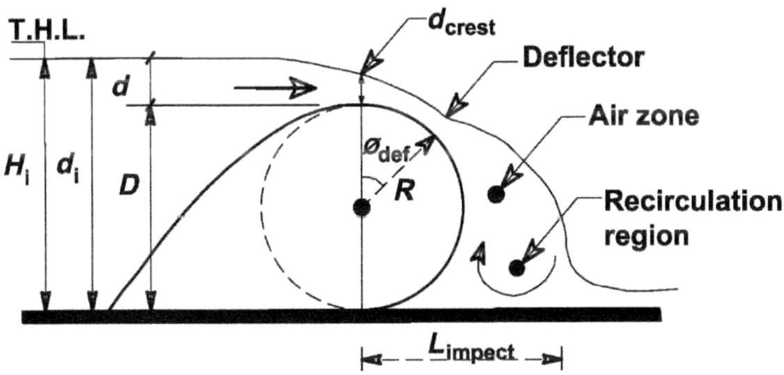

Figure 6.4 Rubber dam overflow (Chanson, 1997).

a new technique for prediction of overflow characteristics, and latest design guiding principle were derived for the most favorable position of the deflector. The nappe adherence on the rubber dam wall might lead to flow unsteadiness at the base of the nappe resulting in pressure fluctuations on the downstream face of the inflated dam and vibrations of the dam material. This can be eliminated by using deflectors as shown in Fig. 6.4.

At any position φ, the dimensionless pressure distribution at the wall equals is as per the following equation:

$$\frac{P_{\text{atm}} - P_s}{\rho_w g R} = \frac{d}{R}\left[F^2 - \cos\phi\left(1 + \frac{d}{2R}\right)\right] \tag{6.1}$$

where d is the depth of flow over crest, P_s is the absolute pressure at the rubber wall, D is the height of inflated dam over the concrete bed, P_{atm} is the atmospheric pressure, ρ_w is the density of water, F is the Froude number, R is the radius of curvature of inflated dam, and ϕ is the angular position with $\phi = 0$ at the crest.

It may be seen from Eq. (6.1), there is an increasing suction pressure ($P_{\text{atm}} - P_s$) down the surface as the flow is accelerated. This may cause the nappe to strict to the dam wall surface and may lead to flow instability and pressure fluctuation on the downstream face of the dam and vibrations of the flexible membrane. Nappe adherence instability can be eliminated by installing a deflector to deflect the nappe off the rubber wall. The resulting Coanda effect acts on the wall surface in direction normal to the flow direction.

The forces acting on the dam under the hydrostatic conditions are

1. internal air or water pressure,
2. upstream water pressure,

3. downstream water pressure,
4. weight of dam material, and
5. wind pressure.

The forces acting on air-inflated and water-inflated dam are shown in Fig. 6.5A and B.

The forces acting on an upstream element and downstream element of the dam per unit length are shown in Fig. 6.6.

The wind pressure on the downstream side, wave formation on upstream water level, and seismic forces are to be taken as per site condition.

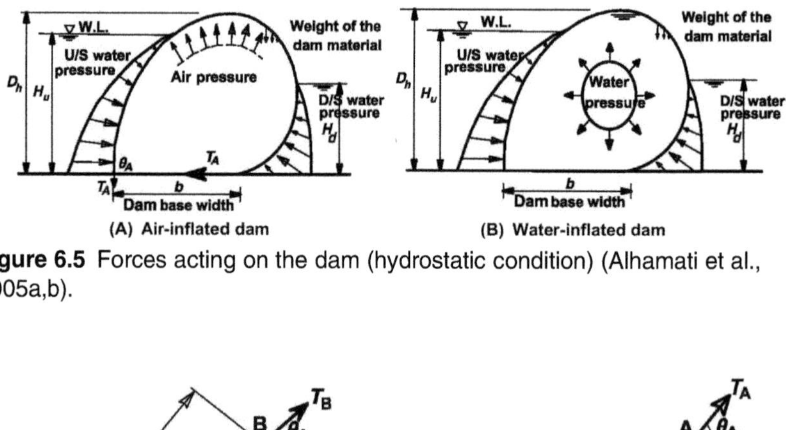

(A) Air-inflated dam (B) Water-inflated dam

Figure 6.5 Forces acting on the dam (hydrostatic condition) (Alhamati et al., 2005a,b).

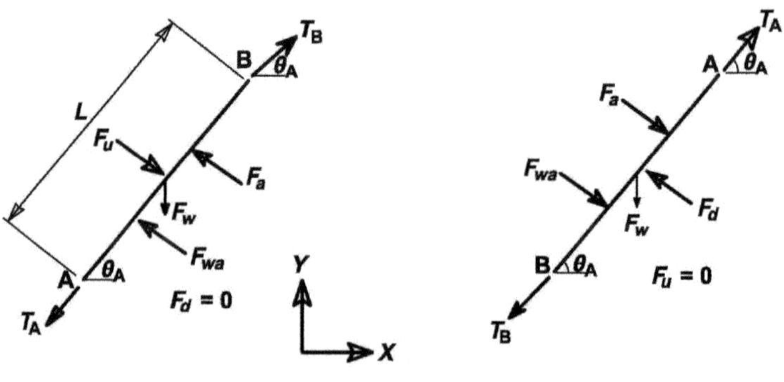

(A) Upstream element (B) Downstream element

Figure 6.6 Forces acting on an element (Alhamati et al., 2005a,b).
Where F_u is the upstream hydrostatic force on the element per unit length, F_d is the downstream hydrostatic force on the element per unit length, F_a is the internal air force on the element per unit length, F_w is the element weight per unit length, F_{wa} is the internal water force on the element per unit length, T_A is the maximum tension force at A of the membrane material per unit length, and T_B is the maximum tension force at B of the membrane material per unit length.

The effect of sediment that is deposited against rubber dam at the river bed level may be important and should be taken into consideration. The effect of dynamic forces caused by water flowing over the dam may require consideration in special cases. During deflation, it is possible for the water pressure to pinch off sections of the dam trapping air in the middle making them impossible to fully deflate. Ice pressure is normally not taken into consideration because ice passes easily by slightly deflating rubber dam.

After deciding of all design parameters, the rubber dam material supplier's carry out the design and analysis of the rubber tube and attached fittings. The stability of the substructure must be evaluated in the same way as any other gravity structure. The manufacturer may provide the force on the anchoring strip for the dam, but if not, all vertical and horizontal forces must be determined as with any other structure.

The forces on the anchoring strip of the dam may be provided by the manufacturer. Vertical and horizontal forces must be determined as with any other structure. External pressure, internal pressure, and weir height are important factors to determine the shape of rubber tube and tensions. The rubber tube attachments fitting should be designed for the tension in rubber fabric and nut tightening forces.

Alhamati et al. (2005a,b) gave Eq. (6.2) for θ_A, slope of the inflatable dam at upstream fixture given as follows:

$$\theta_A = \cos^{-1}\left\{\frac{\left[(\gamma/2) \times \left(H_u^2 - H_d^2\right)\right]}{T_A - 1}\right\} \tag{6.2}$$

where H_u is the upstream water head, H_d is the downstream water head, and T_A is the maximum tension force of the membrane material per unit length.

6.7 Overflow Above and Deflated Flexible Membrane Dam

6.7.1 Introduction

When the rubber dam is deflated, water flows over the dam lined flat on the bed and head loss associated with flat line membrane and damage should be minimum. Deflected rubber membrane can sustain large positive pressure but cannot sustain negative pressure. Some of the flowing hydrodynamic process may lead to erosion, damage or destruction of membrane, hydraulic jump-induced vibration, near-critical flow-induced vibration, wake-induced vibration, and debris erosion.

6.7.2 Overflow of Inflated Rubber Dams

Several researcher indicated vibration problem during overflow about the inflated rubber dam. A summary of model and prototype observation by different researchers have been given in Table 6.1. Considering a fully inflated dam, experimental and computational investigation showed that downstream face of rubber membrane follows closely the shape of circular cylinder.

Further potential instabilities during overflow of rubber dam have been recommended by Chanson (1997) for inflated or deflated position of rubber dam and are given in Table 6.2.

6.7.3 Coefficient of Discharge

Anwar (1967) determined the coefficient of discharge of air and water-inflated dams by using the following equation:

$$C_d = \frac{q}{\sqrt{2}gH^{1.5}} \tag{6.3}$$

where C_d is the coefficient of discharge, q is the discharge over dam per unit length, and H is the overflow head as shown in Fig. 6.7. It is also found that value of C_d lies between 0.35 and 0.5.

Alhamati et al. (2005a,b) determined the formula for calculating (C_d) coefficient of discharge as per the following equation:

$$C_d = 0.5516\left(\frac{H}{D_h}\right)^{0.2697} \tag{6.4}$$

Jumaily and Salih (2005) studied and analyzed performance of air- or water-inflated rubber dams under diverse conditions of inside pressure, upstream, and downstream water heads. Moorthy et al. (1995) applied finite element method to determine the equilibrium shape of the rubber dam and analyzed small vibrations of a pattern of this type. Wu and Plaut (1996) analyzed the vibrations of rubber dams and studied the dynamic behavior of a rubber dam subjected to overflow. Choura (1997) considered the suppression of structural vibrations in air-inflated

Table 6.1 Observations and Recommendations to Prevent Vibration Cases During Overflow Above Inflated Dam (Without Fin) (Chanson, 1997)

Reference	Safe Operation of Rubber Dam	Comments
(1)	(2)	(3)
Anwar (1967)	$\frac{(H_1 - \Delta z_1)}{D} < 1.25$	Laboratory study, 2D model. Air-filled membrane $W = 0.610$ m, $D = 0.229$ m, $\Delta z = 0.076$ m
Shepherd et al. (1969)	$\frac{H_1 - \Delta z_1}{D} < 4$	1:50 laboratory model of Koombooloomba dam spillway, 3D model. Curved spillway crest. Water-filled membrane. Model: $D = 0.0305$ m
Binnie et al. (1973)	$\frac{H_1 - \Delta z_1}{D} < 1.64$	Prototype failure. 3 water-filled membranes. Flat crest $W = 63.7-76.2$ m, $D = 3.05$ m
Ogihara and Maramatsu (1985)	$\frac{H_1}{D} < 1.2$	Laboratory tests. Air-filled membrane $W = 3$ m, $D = 0.2$ m, $\Delta z = 0$
	$\frac{H_1}{D} < 1.3$	Laboratory tests. Water-filled membrane $W = 3$ m, $D = 0.2$ m, $\Delta z = 0$
Takasaki (1989)	$\frac{H_1}{D} < 1.3$	Omata weir model tests. Air-filled membrane $\Delta z = 0$
Economides (1993)	$\frac{H_1}{D} < 1.4$	Review of manufacturers recommendations
Economides and Walker (1994)	$\frac{H_1}{P_{infl}/(\rho_w \times g)} < 1.15$	Laboratory tests. Air-filled membrane Model 1: $W = 0.61$ m, $D = 0.0635$ m (fully inflated), $D_z = -0$ Model 2: $W = 1.83$ m, $D = 0.0635$, $\Delta z = 0$
Chanson (1997)	$\frac{H_1 - D}{R} < \frac{3}{\sigma_c} \times \frac{P_{atm}}{\rho_w \times g \times R} - 6$	Overflow above circular cylinder. Theoretical calculations. Conditions to prevent cavitating separation

Table 6.2 Potential Fluid-Structure Instabilities During Overflow Above Dam (Chanson, 1997)

Mode of Operation	Fluid-Structure Instability	Hydraulic Process	Design/Remedy
(1)	(2)	(3)	(4)
Deflated	Hydraulic jump-induced vibrations	Hydraulic jump above or next to deflated membrane	Avoid deflated dam below or next to hydraulic jump
	Near Critical Flow (NCF)-induced vibrations	Near-critical flow situation in the vicinity of deflated membrane	Avoid dam below or next to near-critical flow
	Wake-induced vibrations	Vortex shedding in wake regions caused by membrane protuberance into the flow and downstream end of membrane	Minimize vortex shedding by streamlining the outer shape of deflated membrane, reducing protuberance into the flow
	Debris erosion	Abrasion by debris overflow and impact by debris	Casing, pneumatic gate, or steel gate. Consider alternative design
Inflated without deflector	Nappe adherence instability	Adherence of overflowing nappe on downstream face of membrane caused by Coanda effect and gravity	Installation of deflector
	NCF-induced vibrations	Near-critical flow situation immediately downstream of dam	Avoid dam below or next to near-critical flow
	Debris erosion	Debris overflow (abrasion, impact)	Consider alternative design

rubber dam by its inside pressure. Hsieh model was employed for devising an approach for vibration control (Hsieh and Plaut, 1990). A dynamic simulation of the response of a rubber dam subjected to a flood was carried out by Lowery and Liapis (1999). Gebhardt (2010) analyzed the vibrations and remedies for water-filled rubber dam and also discussed methods to reduce the vibration.

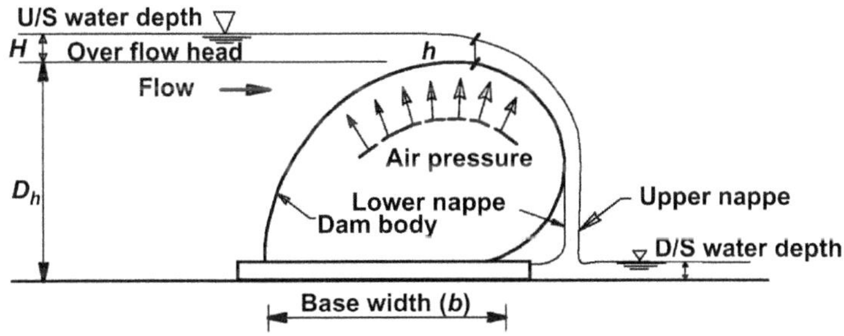

Figure 6.7 Flow over air-inflated dam (Alhamati et al., 2005a,b).

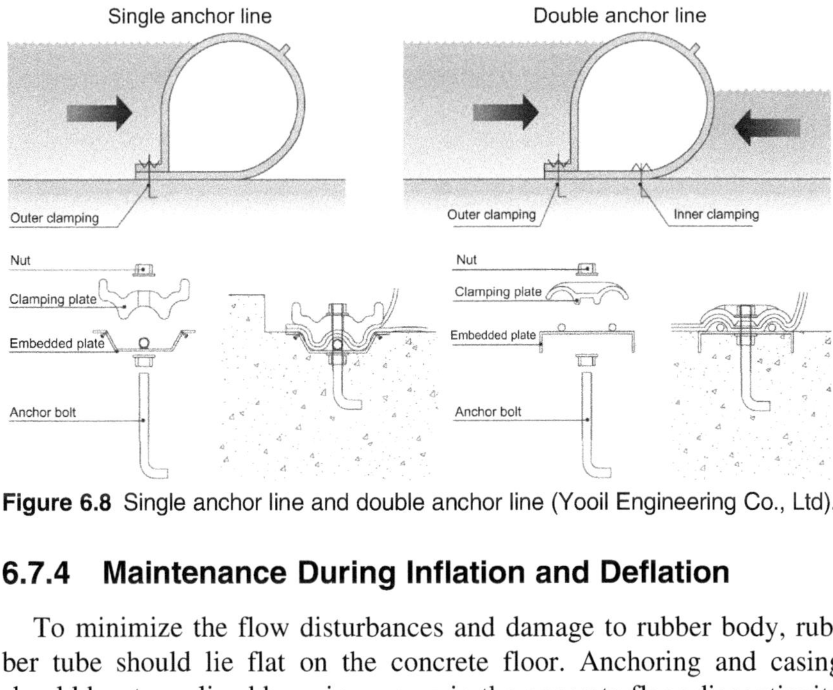

Figure 6.8 Single anchor line and double anchor line (Yooil Engineering Co., Ltd).

6.7.4 Maintenance During Inflation and Deflation

To minimize the flow disturbances and damage to rubber body, rubber tube should lie flat on the concrete floor. Anchoring and casing should be streamlined by using recess in the concrete floor discontinuity due to rubber bag and minimize the vortex shedding effects. The anchor line can be single or double. Single line is suitable for low flow and minimal tail water conditions. Double anchor line can increase stability and reduce the vibration of the dam body and allow higher headwater of overflow. Anchors used for clamping can be made up of from the following materials: stainless steel, cast iron, or carbon material. The single anchor line and double anchor line is shown in Fig. 6.8.

The inflation and deflation property of inflatable weir helps in periodically cleaning of silt and accumulated debris and thus provide high

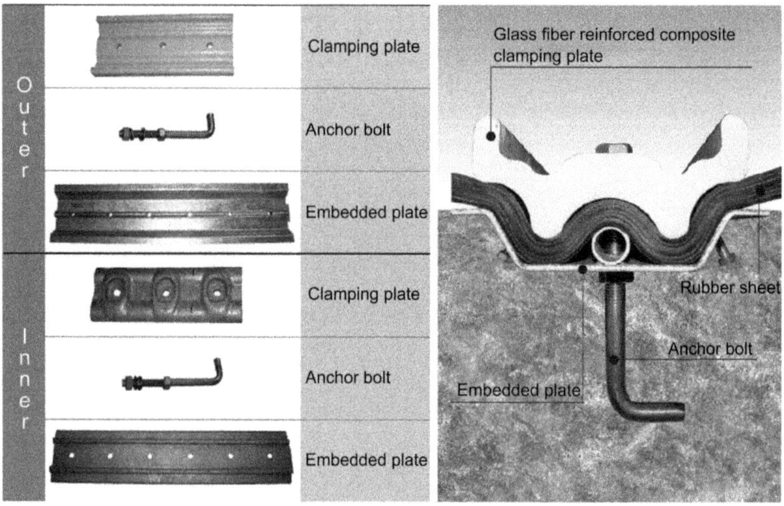

Figure 6.9 Anchor details for rubber dam installation (Yooil Engineering Co., Ltd).

benefit in the abatement of pollution. Anchor details for rubber dam installation are given in Fig. 6.9.

Rubberized fabrics composed of nylon fabric reinforcement for internal strength and coating with protective material like ethylene propylene diene monomer layers of stainless steel meshes and ceramic chips. It is claimed that a coating with steel-mesh reinforcement will prevent fatigue until 1000 cycles of inflation and deflation (Ishimura, 1995). Indian Rubber Manufactures Research Association (Design and Development of Rubber Dams for Watersheds, 2011) had developed twelve diverse types of rubber compounds for cover layer and six diverse types of rubber compounds for cushion latex for the composite material. Eighteen different types of nylon fabrics were designed and developed jointly by Central Institute for Research on Cotton Technology. Out of these, two types were finalized for composite material for rubber dam application. It was evaluated that Vinyl Pyridine latex–Resorcinol-treated nylon fabric shows better adhesion with rubber substrate than styrene butadiene rubber latex–Resorcinol-treated fabric (Jumaily and Salih, 2005).

The magnitude of deformation depends on the type of inflation and air-inflated dam can support a higher upstream water head than a water-inflated dam (Alhamati et al., 2005a,b).

There should be inlet for water to be filled in the inside of rubber dam in case of inflation and outlet for discharging water from the rubber dam in case of deflation. To effect the automatic deflation of an inflatable weir, a water level sensor is installed behind the dam, and

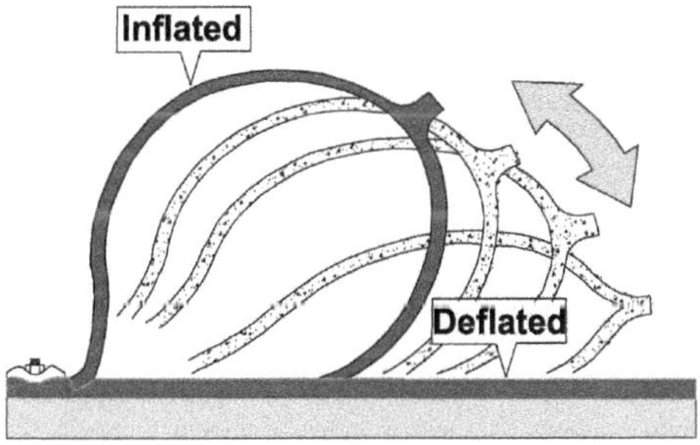

Figure 6.10 Inflation and deflation phenomenon of rubber dam (Sahoo et al., 2010a).

deflation of dam takes place when level of water reaches danger mark. This helps in prevention of damage due to flood. After the flood is over, it can be inflated again. A deflated dam lies flat on the river bottom producing minimal obstruction to river flow. In addition to this, debris and silt is washed out. Flooding debris and sediment tend to accumulate behind a conventional weir. During deflation of an inflatable weir, when the water is released, sediments and debris are washed out. This cleansing action provides a mechanism for removal of entire sediment deposit upstream of inflatable weir. By partial deflation, routine cleaning can also be done. Routine cleansing of the weir can be carried out by partial deflation. Rubber dam eliminates the sediment and piled soil in both up and downstream when deflated. In operating an inflatable weir, its performance should be determined regarding the weir safety and also the intake function and to assure the absence of adverse influences on both the upstream and downstream of river. As pumping/withdrawal of large amount of air/water is involved in the inflation/deflation process, the time taken in the process can be longer and may influence the rate of raising/lowering of the river water levels, which ultimately affects the weir performance in accomplishing the objectives (Khatsuria, 2004).

The inflation and deflation phenomenon of rubber dam is shown in Fig. 6.10. The inflation and deflation procedure of both air- and water-deflated rubber dam is same (Zhang et al., 2002).

6.8 Design Criteria for Rubber Dam System

Rubber dam shall be designed and installed to meet following operational requirement.

6.8.1 Inflation

1. Upstream side should be inspected to ensure that by raising the water level, people and property should not be adversely affected.
2. The debris (especially sharp objects) should be removed which are near to rubber dam body.
3. Air/water exhaust valve is closed.
4. Air/water supply valve is opened.
5. Air compressor/pump is started.
6. Air/water pressure gauge is monitored during inflation process. Adjust the air pressure according to the headwater level. If the headwater level is less than half of the dam height, the air pressure should be approximately 60% of the design pressure. The air pressure should increase in proportion to the headwater level. Increase the air pressure to the design level after the headwater reaches the design level. Then turn off the air compressor and close the air supply valve.
7. Fill rate must be able to be varied to allow for the flow in the flood control channel to flow past the dam and/or to overtop the partially inflated dam for the first 2 hours until the dam is fully inflated.
8. Full inflation shall be determined by the inner pressure in the rubber dam bladder and the required operating level below.
9. Rubber dam system supplier/manufacturer (RDSS) shall determine and identify other governing factors which may impact rubber dam inflation rate.

6.8.2 Deflation

1. Downstream side should be inspected to ensure that by raising the water level, people, and property should not be adversely affected.
2. The debris (especially sharp objects) should be removed on the downstream side of the dam foundation, because they may damage the rubber dam fabric during deflation process.

3. Drainpipe/air exhaust value is opened to drain the water/ air.
4. Air/water exhaust valve is opened to deflate the dam.
5. Full deflation shall be determined by the inner pressure in the rubber dam bladder.

6.8.3 Emergency Mechanical Deflation System

1. The RDSS shall design, manufacture, and supply a mechanical deflation system. The system shall allow partial deflation of the rubber dam bladder when internal pressures in the bladder exceed the calculated maximum design limits and until the pressure in the rubber dam bladder is reduced to acceptable levels.
2. The design of the emergency mechanical deflation system shall include all equipment required, pipes, flanges, floats, valves, etc. to guarantee that the system will operate to relieve the excess pressure without the use of any external power.
3. The emergency mechanical deflation system shall be designed to fit within the constraints of the existing rubber dam control building.

6.8.4 Design Flow Event With Rubber Dam Bladder Deflated

The design flow event will be equal to the 1 in X year design flood event in the flood control channel where the rubber dam bladder is installed. X is the return period depending upon the project. The peak discharge should be calculated and also maximum expected velocity of the water. The probability of occurrence of event (say the maximum flood discharge observed or likely to occur in a year at a location on a river), whose magnitude is equal to or in excess of a specified magnitude X is denoted by P. A related term, the recurrence interval (also known as the return period), is defined as $T = 1/P$. This represents the average interval between the occurrence of a flood peak of magnitude equal to or greater than X.

6.8.5 Rubber Dam Bladder Specification

1. General:
   a. The rubber used to fabricate the bladder shall have superior weathering properties with regard to prolonged exposure to ultraviolet rays and ozone.
   b. All components or accessories to be fixed or fitted inside the rubber bladder shall be rounded or smooth with no sharp edges.
2. Operational temperature range—28–115°F
3. Operational humidity range—10%–100%
4. Overflow conditions to be considered:
   a. The bladder will not be exposed to continuous overflowing water.
   b. The majority of time the bladder will be inflated with no water overtopping the bladder.
   c. When the bladder is deflated, the available flows in the flood control channel will flow over the bladder.

6.8.6 Rubber Dam Bladder Anchorage

1. General:
   a. All metal components of the rubber dam system, including but not limited to anchor bolts, clamp plates, base plates, or embedded plates, metallic piping, pipe flanges, and select portions of the deflation equipment, shall be protected against corrosion by hot dip galvanizing unless specified otherwise in other parts of this specification.
   b. If the parts are fabricated using stainless steel, they shall not be hot-dipped galvanized.
2. Concrete foundation: The new rubber dam system shall be installed on a concrete foundation.
3. Special pipe connections to rubber dam bladder:
   a. Water supply and drainpipe flanges and bladder Interface—Pipe material and diameter to be designed by RDSS to meet the required operational objectives.
   b. Pressure sensing pipe flanges—Pipe diameter and material to be designed by RDSS to meet the operational objectives.

6.8.7 Operational Equipment

1. Water fill pump:
   a. The fill pump shall be sized, flow rate, and required pressure to be capable of filling the rubber dam bladder within the time required in rubber dam inflation rate.
   b. Suction and discharge head and volumetric capacity of the fill pump shall be determined.
   c. Pumps and motors used for the water fill pump system.
2. Piping:
   a. Design the water fill pump suction, discharge, and drain line pipe layouts required for all operating conditions of the rubber dam. It is preferred that the pipe connections and portions of the existing pipe to and from the existing rubber dam control building be reused, if possible.
   b. The pipes shall be sized to ensure that the rubber dam bladder can be inflated and deflated within the time frames required.
   c. Flanged joints or other approved joints shall be included in the design at suitable locations to facilitate ease of dismantling of the piping system for maintenance.
3. Control equipment:
   a. Design the stand-alone control equipment required by this contract to operate the rubber dam system as described in this specification.
   b. The control equipment must be designed to measure and record the dam crest elevation, the inner pressure of the rubber dam bladder, the upstream impound water level, and other relevant parameters as required to safely operate the rubber dam system in as safe a manner as possible.

In case of water-inflated rubber dam, control shaft system consists of following four chambers.

1. *Pump shaft*: There are inlet pipes from the forebay into the pump shaft equipped with shut off valves. The pump shaft can be drained by the valve any time. Pumps were situated in this shaft to supply filling water to the filling shaft through the pipe work.

2. *Filling shaft*: The filling shaft is equipped with manual sluices to close or open the filling pipes to the membrane with a probe.
3. *Control shaft*: Control shaft consists of probe, manual sluices for the drainage pipe, drainage pump is equipped to deflate the rubber dam with normal speed.
4. *Drainage shaft*: Drainage shaft contains the main accoutrements for drainage of the control shaft and the membrane, respectively.
   a. Equipment safety:
      i. The rubber dam bladder shall deflate automatically and completely when water depth over the fully inflated rubber dam bladder crest exceeds 0.6 m.
      ii. The rubber dam bladder shall automatically partially deflate when the inner pressure exceeds a predetermined value. This internal pressure limiting value shall be documented in the project operation and maintenance and include a discussion of how it was determined.
      iii. The water supply and drain system shall be designed, so that the rubber dam bladder cannot be overpressured if the water pump fails to stop. A pressure-relief valve shall be installed on the water supply pipe for this purpose. The size of the pressure-relief valve shall be sufficient to divert the full flow produced by the water fill pump and return it to Shinn pond.
      iv. A mechanical pressure-relief device such as a siphon pipe, as described above, shall be provided to prevent overpressure of the rubber dam bladder.
   b. When the rubber dam bladder is fully inflated:
      i. The rubber dam bladder must maintain a uniform dam crest elevation while in operation.
      ii. Develop a dam weir crest flow algorithm during design of the rubber dam system. This weir crest flow algorithm shall be submitted to the project engineer prior to commissioning of the rubber dam system and will be used during the commissioning of the rubber dam system.

c. When the rubber dam bladder is fully deflated:
   i. The rubber dam bladder shall lie flat on the existing concrete foundation or down into the constructed plunge pool on the downstream side of the rubber dam foundation.
   ii. The rubber dam bladder shall not float or move under any flow conditions when deflated.

6.8.8 Rubber Dam Bladder, Anchoring, and Pipe Connections

The rubber dam shall be designed to withstand the maximum forces imposed on the rubber bag under the highest specified upstream and downstream water levels. The rubber dam shall be designed to be stable in all specified operating conditions, which will be determined during design. The design shall mitigate potentially harmful vibrations of the fully inflated rubber body that may occur during normal operation. The design of the control and safety system shall guarantee positive deflation of the rubber body before the upstream water reaches the level at which damaging vibrations can occur.

The rubber dam shall be designed for frequent and regular operation.

1. A multitime per weekly inflation/deflation cycle shall not be reason for overstressing or premature wear of any of the rubber dam components.
2. The rubber dam shall also be capable of being operated in the fully inflated or fully deflated positions for extended periods without any harmful or detrimental effects to the rubber dam or its related components.

6.9 Seismic National and International Requirements

The rubber dam system shall be designed to resist shear, bending, and overturning moments along the three orthogonal axes.

6.10 Operational Steps for Water-Filled Dam

The inlet pumps are to be operated from control shaft to lift the rubber dam. Care should be taken that the water on the upstream should not be

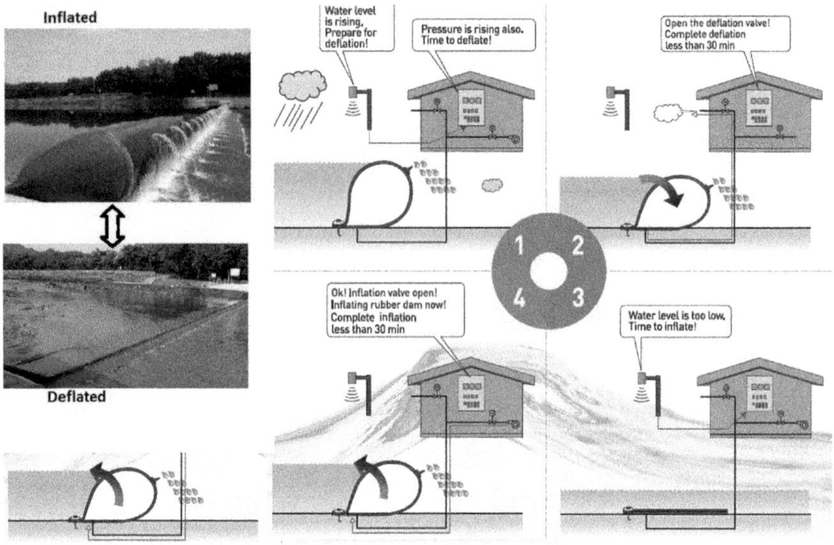

Figure 6.11 Operation procedure of rubber dam (Yooil Engineering Co., Ltd).

less than half a meter from top of the rubber. In the control room, the rubber dam level and water levels will be automatically screened and regulated accordingly. If the inflows are lean, proper time should be taken to pool the water on U/S side, and the rubber dam is to be raised to the maximum height duly observing water level not less than half a meter. For lowering the dam, the drainage pipe is to be operated.

The drainage pipe can be operated both manually and electrically to facilitate lower down the rubber foil during floods operate the air supplier in the operation room with fully automatic, semiautomatic, or manual to supply the rubber dam with air up to the specified or desired pressure and height. The ultrasonic sensor is used for detecting inflation and deflation water level. The system will open the exhaust valve by either the flood level or your necessity to discharge the air inside of the rubber dam. When the air is fully exhausted, the rubber dam adheres closely to the bottom of a river thus, maximizing the water discharge within the area (Sahoo et al., 2010a,b). Fig. 6.11 shows different operation processes of rubber dam.

6.11 Inflation Time

Inflation time depends on three factors: the inner volume of the dam body, the external pressure acting on the dam body, and the capacity of the air compressor or water pump. The time required for full inflation/deflation can be varied depending on the application. There should be inlet for water to be filled in the inside of rubber dam in case of inflation and outlet for discharging water from the rubber dam in case of deflation.

6.12 Optimization of Suitable Inflation Deflation Mechanism

Optimization of suitable inflation deflation devices depends on availability, size, and thickness of rubber dam fabric, maximum height of inflated weir, and minimum inflation and deflation time. If the rubber dam is for irrigation purpose for farmers, then simple user-friendly technique should be used, so that it will be easy for farmers to operate (Sahoo et al., 2010b). Fig. 6.12 shows the farmer-friendly operation of rubber dam for agricultural purpose.

6.13 Safe and User-Friendly Operation

Operate the air supplier in the operation room with fully automatic, semiautomatic, or manual to supply the rubber dam with air up to the specified or desired pressure and height. The ultrasonic sensor is used for detecting inflation and deflation water level. The system will open

Figure 6.12 Chandeshwar-I rubber dam Kurda Orissa, India (Sept 2013 site visit).

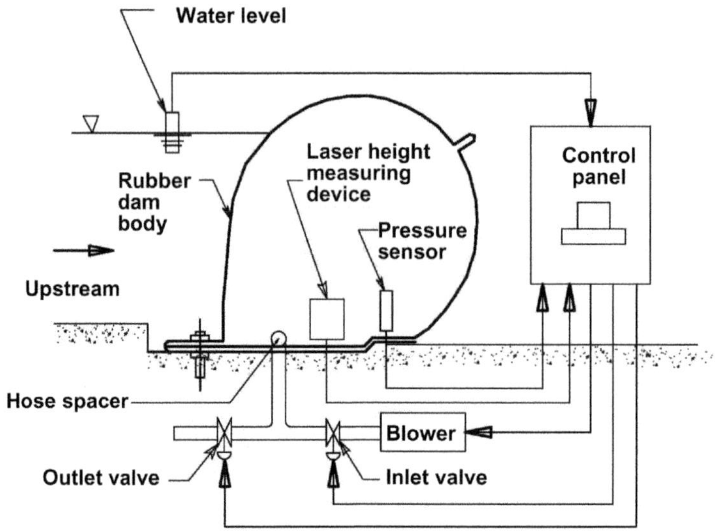

Figure 6.13 Automatic control mechanism of rubber dam (Zhang et al., 2002).

the exhaust valve by either the flood level or your necessity to discharge the air inside of the rubber dam. When the air is fully exhausted, the rubber dam adheres closely to the bottom of a river thus, maximizing the water discharge within the area. Rubber dams can be set to deflate at a particular level. Once that level is reached, deflation mechanism is triggered and deflation takes place. A pressure sensor as shown in Fig. 6.13 is used to observe the pressure of the inflated rubber body. The inlet of air/water to the rubber dam body is cut off automatically when inner pressure reaches the present level. This prevents damage to the dam body due to extreme pressure. The internal pressure of an air-filled rubber dam can be maintained at a steady value even if the rubber dam is installed in a changeable external environment. A laser height measuring can be installed within the dam body to constantly measure its height when during inflation or deflation (Zhang et al., 2002).

Fig. 6.14 shows the sediment discharge system. Install water intake pipes on the river bottom at short intervals to vacuum in the sediments and water pollutants with bottom water by using its own water pressure.

Rubber dams help in flood control due to the automatic property of inflation and deflation (Tam and Zhang, 1999). Annual cost and life cycle cost of inflated weir is less than other conventional types of weirs (Ul Islam and Kumar, 2015, 2016).

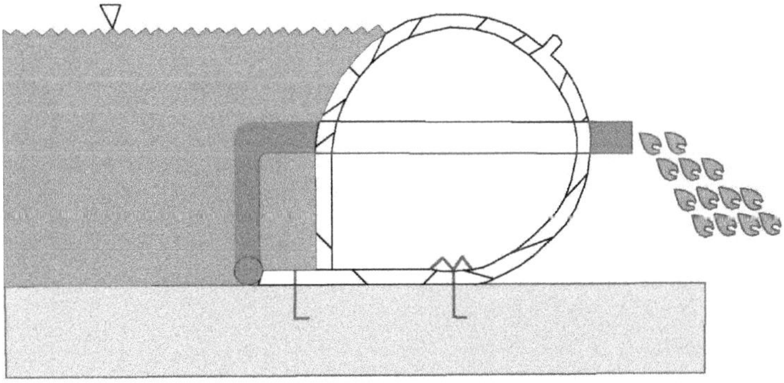

Figure 6.14 Sediment discharge system of rubber dam (Yooil Engineering Co., Ltd).

6.14 Reference Standards

The following standards may be applied for design, manufacturing, supply, installation, and operation of rubber dams.

1. American Society of Mechanical Engineers (ASME):
   a. Section IX of the ASME Boiler and Pressure Vessel Code
   b. ASME B18.21.1-2009—Washers—Helical Spring-Lock, Tooth Lock, and Plain Washers (Inch Series)
   c. ASME A666-10—Standard Specification for Annealed or Cold Worked Austenitic Stainless Steel Sheet, Strip, Plate, or Flat Bar
2. American Society for Testing and Materials (ASTM):
   a. ASTM A36/A36M-12—Standard Specification for Carbon Structural Steel
   b. ASTM A123—Standard Specification for Zinc (Hot Dip Galvanized) Coatings on Iron and Steel Products
   c. ASTM A153—Standard Specification for Zinc Coatings (Hot Dip) on Iron and Steel Hardware
   d. ASTM A193/A193M—12b Standard Specification for Alloy Steel and Stainless Steel Bolting for High Temperature or High Pressure Service and Other Special Purpose Applications
   e. ASTM A194/A194M—12a Standard Specification for Carbon and Alloy Steel Nuts for Bolts for High Pressure or High Temperature Service, or Both

  f. ASTM A240/A240M—13c Standard Specification for Chromium and Chromium-Nickel Stainless Steel Plate, Sheet, and Strip for Pressure Vessels and for General Applications
  g. ASTM A354-11—Standard Specification for Quenched and Tempered Alloy Steel Bolts, Studs and Other Externally Threaded Fasteners.
  h. ASTM F436M-11—Standard Specification for Hardened Steel Washers (Metric).
  i. ASTM A536—Standard Specification for Ductile Iron Castings.
  j. ASTM A563—Standard Specification for Carbon and Alloy Steel Nuts.
  k. ASTM F1554—Standard Specification for Anchor Bolts, Steel, 36, 55, and 105 ksi Yield Strength.
3. American Welding Society (AWS): AWS D1.1—Structural Welding Code Steel.
4. German Institute for Standardization (DIN)
  a. IN EN ISO 252—Conveyor Belts—Adhesion between Constitutive Elements – Test Methods.
  b. DIN ISO 1817:2005—Rubber, Vulcanized, or Thermoplastic—Determination of the effects of Liquids.
  c. DIN 53504-S2—Testing of Rubber—Determination of Tensile Strength at Break; Tensile Strength at Yield, Elongation at Break and Stress Values in a Tensile Test.
  d. DIN 53505—Testing of Rubber—Shore A and Shore D Hardness Test.
  e. DIN 53508—Accelerated Aging of Rubber.
5. International Organization for Standardization (ISO).
  a. ISO 34-1:2010—Rubber, Vulcanized or Thermoplastic—Determination of Tear Strength—Part 1; Trouser, Angel & Crescent Test Piece
  b. ISO 37:2011—Rubber, Vulcanized or Thermoplastic—Determination of Tensile Stress-Strain Properties
  c. ISO 283:2007—Textile Conveyor Belts—Full Thickness Tensile Strength, Elongation at Break and Elongation at Reference Load

d. ISO 815-1:2008—Rubber, Vulcanized or Thermoplastic—Determination of Compression Set—Part 1
e. ISO 1183-1:2012—Methods for Determining the Density of Non-Cellular Plastics (Methods A, B, and C)
f. ISO 1431-1:2012—Resistance to Ozone Cracking—Part 1: Static and Dynamic Strain Testing
g. ISO 1461:2009—Hot Dipped Galvanized Coatings on Fabricated Iron & Steel Articles—Specifications and Test Methods
h. ISO 3506-1:2009—Mechanical Properties of Corrosion-Resistant Stainless Steel Fasteners—Part 1: Bolts, Screws, and Studs
i. ISO 4032:2012—Hexagon Regular Nuts (Style 1)—Product Grades A & B
j. ISO 4649:2010—Rubber, Vulcanized or Thermoplastic – Abrasion Resistant using Rotating Cylindrical Drum Device
k. ISO 7089:2000—Plain Washers—Normal Series—Product Grade A
l. ISO 7090:2000—Plain Washers, Chamfered—Normal Series—Product Grade A
m. ISO 8673—Hexagon Regular Nuts (Style 1)—With Metric Fine Pitch Thread—Product Grades A & B

6. Japanese Technical Standard of Rubberized Fabric Inflatable Weirs, Second Draft, August 1983.

6.15 Conclusion

Inflatable weirs are easy to inflate and deflate resulting in passage of sediments and boulders easily from upstream to downstream of rubber dam due to which damage due to flood is reduced. Thus, the cost of operation and maintenance, repair works, sediment removal, and flood damage is minimal. During deflation of rubber dam, puncturing and wear and tear can take place because of sharp objects and debris. This damage can be minimized by routinely cleaning the river beds in upstream as well as in downstream of the inflated weir and also by using thicker dam body or a ceramic-chip-coated rubber dam. To safeguard the dam body against punctures by waterborne debris, polyethylene-based foam blocks may be placed inside the rubber dam body to act as a cushion. The use of cut-resistant rubber is a recent innovation. Both ceramic-chip-coated rubber dams and stainless steel-mesh reinforced rubber dams are now available in the market. Many rubber dams in the United States have been coated with ceramic chips for protection against vandals.

References

Alhamati, A.A., Mohammed, T.A., Ghazali, A.H., Norzaie, J., Jumaily, K.K., 2005a. Determination of coefficient of discharge for air inflated dam using physical model, Suranaree. J. Sci. Technol. 12 (1), 19–27.

Alhamati, A.A., Mohammed, T.A., Norzaie, J., Ghazali, A.H., Jumaily, K.K., 2005b. Behaviour of inflatable dams under hydrostatic conditions, Suranaree. J. Sci. Technol. 12 (1), 1–18.

Anwar, H.O., 1967. Inflatable dams. J. Hydraul. Div. ASCE 93 (HY3), 99–119.

Binnie, G.M., Thomas, A.R., Gwyther, J.R., 1973. Inflatable weir used during construction of Mangla dam. Proc. Inst. Civ. Eng. Part 154, 625–639.

Chanson, H., 1997. A Review of the overflow of inflatable flexible membrane dams, Australian Civil Engineering Transactions, Vol. 39, No. 2/3, pp. 107–116. <https://www.search.informit.com.au/documentSummary;dn = 201267076554107;res = IELENG> ISSN: 0819-0259. [cited 08 Nov 18]

Chanson, H., 1998. Hydraulics of rubber dam overflow: a simple design approach. In: Thompson, M.C., Hourigan K. (Eds.), Proceedings of the 13th Australasian Fluid Mech. Conf., Melbourne, Australia, 13–18 December, vol. 1, pp. 255–258.

Choura, S., 1997. Suppression of structural vibrations of an air-inflated membrane dam by its internal pressure. J. Comput. Struct. 65 (5), 669–677.

Design and Development of Rubber Dams for Watersheds, Half Yearly Progress Report (April 2011–September 2011) published by Indian Rubber Manufacturers Research Association (IRMRA), Mumbai.

Economides, T.A., 1993. Experimental investigation on the dynamics of inflatable dams (Dissertation). Virginia Tech.

Economides, T.A., Walker, D.A., 1994. Non-intrusive experimental setup for inflatable dam models. Fundamentals and Advancements in Hydraulic Measurements and Experimentation. ASCE.

Gebhardt, M., On the causes of vibrations and the effects of countermeasures at water-filled inflatable dams. In: IAHR-2010-European Congress. May, 2010

Hasan, M.S., Kabir, M.I., 2014. Feasibility of Natore Rubber Dam on Mahananda River in Bangladesh and its performance on irrigation. Am. J. Eng. Res. 3 (1), 27–34.

Hitch, N.M., Narayanan, R., 1983. Flexible dams inflated by water. J. Hydraul. Eng. 109 (7), ASCE, ISSN 0733-9429/83/0007-1044, pp. 1044–1048.

Hsieh, J.-C., Plaut, R.H., 1990. Free vibrations of inflatable dams. Acta Mech 85, 207–220.

Imbertson, N.M., 1960. Automatic rubber diversion dam in the Los Angeles River. J. Am. Water Works Assoc. 52, 1373–1378.

Ishimura, Y., 1995. Design and installation of inflatable rubber weir. In: International R and D Conference, CBIP, New Delhi, India.

Jumaily, K.K., Salih, A.A., 2005. Analysis of inflatable dams under hydrostatic conditions. J. Eng. Dev. 9 (3), ISSN 1813-7822, 69–92.

Khatsuria, R.M., 2004. Chapter 15. Inflatable Rubber Weirs of Hydraulics of Spillways and Energy Dissipators. CRC Press, ISBN 0203996984.

Lowery, K., Liapis, S., 1999. Dynamic analysis of an inflatable dam subjected to a flood. Comput. Mech. 24, 52 $\pm$ 64 Ó Springer-Verlag 1999, 52–64.

Moorthy, C.M., Reddy, J.N., Plaut, R.H., 1995. Three dimensional vibrations of inflatable dams, Thin-Walled Structures, vol. 21. Elsevier Science Ltd, Amsterdam, pp. 291–306.

Ogihara, K., Maramatsu T., 1985. Rubber dam: causes of oscillation of rubber dams and countermeasures. In: Proceedings of the 21st IAHR Congress, Melbourne, Australia. pp. 600–604.

Rantz, S.E. et al., 1982. Measurement and computation of streamflow: volume 2. Computation of discharge. Geological Survey Water-Supply Paper 2175. United States Government Printing Office, Washington.

Tam, P.W.M., Zhang, X.-Q., 1999. Management of rubber dams in Hong Kong. Can. J. Civil Eng. 26 (2), 123–134.

Yooil Engineering Co., Ltd. Rubber Dam Report. Yooil Engineering Co., Ltd. <www.yooileng.co.kr/eng/> (accessed 27.05.18).

Sahoo, N., Jena, S.K., Kumar, A., Mishra, A., Patil, D.U., Anandl, P.S.B., 2010a. Design, development and installation of indigenous rubber check dam in watershed. Proceedings of All India Seminar on "Engineering Interventions to Enhance Income of Small and Marginal Farmers". Institute of Engineers, New Delhi, pp. 183–191.

Sahoo, N., Jena, S.K., Kumar, A., Mishra A., Anand, P.S.B., 2010b. Rubber dams—a new concept for watershed. In: Proceedings of the XLIV ISAE Annual Convention and Symposium on Engineering Interventions for food security, New Delhi, 28–30 January 2010, p. 3.42.

Sarker, K.K., Wang, X.Y., Islam, N., Xu, C.L., Qiao, X.D., 2011. Performance evaluation of the rubber dam project for irrigation development. Sci. Res. Essays 6 (22), 4700–4707. Available from: http://www.academicjournals.org/SRE.

Shepherd, E.M., McKay, F.A., Hodgens, V.T., 1969. The Fabridam extension on Koombooloomba dam of the Tully falls hydro-electric power project. J. Inst. Eng. Aust. 41, 1–7.

Takasaki, M., 1989. Omata inflatable weir, at the Kawarabi hydro scheme, Japan. Int. Water Power Dam Constr. 41 (11), 39–41.

U.S. Army Corps of Engineers, 2001. Ice engineering, performance survey of inflatable dams in ice-affected waters. U.S. Army Research and Engineering Laboratory, Ice Engineering Information Exchange Bulletins. Hanover, New Hampshire, pp. 1–5.

Ul Islam, S., Kumar, A., 2015. Evaluation of rubber dams for SHP in India. Indian J. Sci. Technol. 8 (28).

Ul Islam, S., Kumar, A., 2016. Inflatable dams for SHP projects. Renew. Sustain. Energy Rev. 57, 945–952. Available from: http://www.indjst.org/index.php/indjst/article/view/84090.

Watson, R., 1985. A note on the shapes of flexible dams. J. Hydraul. Res. 23 (2), 179–194.

Wu, P.H., Plaut, R.H., 1996. Analysis of the vibrations of inflatable dams under overflow conditions. Thin-Walled Struct. 26 (4), 241–259.

Zhang, X.Q., 2007. Flexible structures. Int. Water Power Dam Constr. 59 (2), 32–37.

Zhang, X.Q., Tam, P.W.M., Zheng, W., 2002. Construction, operation, and maintenance of rubber dams. Can. J. Civil Eng. 29, 409–420.

7 Life Cycle Assessment of Rubber Dam

***Shahid ul Islam*[1] *and Arun Kumar*[2]**

[1]Baba Ghulam Shah Badshah University, Rajouri, India, [2]Alternate Hydro Energy Centre, Indian Institute of Technology, Roorkee, India

Abstract

Diversion heads are constructed in a river to divert the water to ensure a continuous supply of sediment-free water diverted into the canal. This diverted water can be used for various purposes viz., hydropower generation, water resource management, irrigation, flood control and fisheries. The conventional weirs, including bush and boulder weir, trench weir (tyrolean), raised gravity weir, and "*mathu bund*" (a local name), create afflux at the time of floods which result in submergence. Conventional weirs take long time to construct, and also operation and maintenance costs are more because they create obstruction to the passage of flood. Rubber dam is selected based on the design parameters for the particular sites. Life cycle costs (LCC) of different types of weirs, including inflatable weir and their comparison, are presented. For life cycle costing, all the costs viz., capital cost, operation, and maintenance cost associated with project are added. The annual cost and LCC were worked out using engineering designs and drawings, and taking into account the cost of operation and maintenance as well as different losses such as head loss, water loss, and repairs. Damage due to major floods for different types of weir is also taken into account by monetizing the loss of power.

Keywords: Life cycle; trench weir; rubber dam; greenhouse gas; mathu bund; weir

7.1 Introduction

Diversion headworks are constructed in a river to provide head divert the water so that continuous supply of silt-free water can be diverted in

Hydraulic Rubber Dam. DOI: https://doi.org/10.1016/B978-0-12-812210-5.00008-0

the canal. Headwork may be having diurnal or no storage. In diurnal storage headwork, water is stored when it is available in surplus and released when demand is more than supply during the day. A weir is constructed across the river to supply water to the off taking channel. The diverted water can be used for various purposes viz., hydropower generation, water resource management, irrigation, flood control and fisheries, etc.

The impacts of concrete-based weirs are different in different phases of construction, operation, and decommissioning. This impact can be minimized by identifying possible impacts, then calculating the risk and minimizing the risk by adopting various techniques that could help to lower the overall impacts. The major hazards during the construction phase of weir which impact the environment and society include demand for construction material, excavation process, and vehicles. The equipment and vehicles engaged in construction phase mostly draw energy from fossil fuels and emit greenhouse gases. During the construction of Glen Canyon dam, the CO_2 emission was estimated as 8×10^5 MT (Pacca and Horvath, 2002).

During the construction phase, there is also risk of accidents to the workers. The site selected for construction purpose may lead to migration and disappearance of some animals and plants forever. During the operation phase, weirs and dams also pose threat to the environment and people to some extent. During the operation phase, if the weir/dam fails, it may cause loss of life and damage to property. During operation phase, the decay of biomass emits CO_2 in the reservoir and estimated emission of CO_2 equivalent from the Glen Canyon dam was 3.5×10^5 MT (Pacca and Horvath, 2002). The most destructive event during the life cycle is the failure of dam or weir. During dam failure, the dam can be fully or partially damaged (Stewart, 2000). Warning sign and hooters dam help to reduce loss of life and property (Emergency Management Ontario EMO, 2005).

The last stage of the life cycle is the decommissioning phase which poses threat of flooding the downstream side of dam thereby causing loss of lives and property. During this stage, trapped compressed CO_2 is emitted as compared with other two phases. The emission during this phase from Glen Canyon dam was estimated as 3.3×10^6 MT of CO_2 equivalent (Pacca and Horvath, 2002). The loss of lives and property in this phase is less than dam failure as it is done in controlled manner. The environmental and social consequences are there throughout in a dam's life (Jackson and Sleigh, 2000).

The dust released during construction phase also contributes to air pollution. Due to the construction of dam or weir, silt carrying capacity of river is reduced as it provides a silent pool of water in the form of reservoir in which silt and sediments are deposited. Watershed management which includes plantation, landslide, and erosion control in the upstream face of dam or weir will help to reduce the sediment load on reservoir. The deposition of sediments in the reservoir reduces the storage capacity of the reservoir and periodic cleaning of dams or weirs can help to reclaim the capacity to its original form. These sediments deposited inside the reservoir also contain carbon which is released during decommissioning phase of dam or weir.

The conventional weirs create afflux at the time of floods which result in submergence. Due to the high level of crest above the river bed, large amount of silt remains without getting washed-out downstream. Conventional weirs take long duration of time for construction and also operation and maintenance cost is more because they create obstruction to the passage of flood. Opening and closing of shutters/ gates is time consuming and requires considerable labor and electricity. Rubber dam is selected based on the design parameters for the particular site. Design parameters include membrane length, anchorage distance, and inflation pressure (Islam and Kumar, 2015, 2016). In hilly areas the conventional weirs are damaged because steep surface cause boulder movement at the time of floods (Rickard et al., 2003). The main application of inflatable dams is to divert water for hydropower and irrigation. The inflatable dam is also used in water resource management for flood control and to raise the height of existing weirs and spillways. Inflatable weirs are easy to inflate and deflate resulting in passage of sediments and boulders easily from upstream to downstream of rubber dam due to which damage due to flood is reduced. Inflatable dams are a feasible alternative to return river to its natural condition which is recent trend.

The rubber dam has been used for different applications such as irrigation, water supply, power generation, and recreation worldwide being flexible, simple in structure and reliable. The inflation and deflation property of inflatable weir helps in routinely cleaning the silts and provides safe passage during floods. In this chapter, life cycle cost (LCC) of different types of weirs including inflatable weir and their comparison is presented. For life cycle costing, all the costs viz., capital cost, operation, and maintenance cost associated with project are added. The life cycle costing of different weirs has been carried at five different sites as case study. The conventional weirs included bush and boulder weir, trench weir (Tyrolean), raised gravity weir, and "mathu bund" (a local name) have been compared with rubber dam. The annual cost and LCC was worked out using engineering designs and drawings and taking into account the cost of operation and maintenance as well as different losses on account of head loss, water loss, and repairs. Damage due to major floods for different types of weir has been also taken into account by monetizing the loss of power.

7.2 Life Cycle Assessment

Life cycle assessment (LCA) is a systematic tool for analyzing and evaluating the cost and impacts over the entire life cycle of a product or system or service from "cradle to grave" (Jensen, 1997). LCA involves tracing of all the major stages and processes involved over the life cycle of a product or system or service covering. The impact due to extraction of raw materials, manufacturing, product use, recycling, and final disposal needs to be identified and quantified at each stage. LCA facilitates a systems view in impact, cost and benefit evaluation of products, materials, and processes. LCA attempts to quantify cost over the entire life cycle of a product from raw material extraction, manufacturing, and use to ultimate disposal. Life cycle is the consecutive and interlinked stages of a product or service system, from the natural resource extraction to its final disposal. The LCC of an asset is defined as "the total cost throughout its life including planning, design, acquisition and support costs and any other costs directly attributable to owning or using the asset." LCA is a systematic set of procedures for compiling and examining the inputs and outputs of materials and energy and the associated environmental impacts directly attributable to the functioning of a product or service system throughout its life cycle.

There are four linked components of LCA.

- *Goal definition and scoping*: the purpose of LCA is identified, boundaries are determined, and goal is defined.
- *Life cycle inventory*: the energy and raw material usage and environmental releases at every stage of production are quantified.
- *Impact analysis*: impact assessment on health of humans and environmental impact assessment is done.
- *Improvement analysis*: different opportunities to reduce energy, material inputs, or environmental impacts at each stage of the product life cycle are evaluated.

In most of the studies on LCA method, construction phase is widely considered and biomass decay has been considered by non-life cycle studies. Limited studies have been carried regarding the combined effect of construction, operation, and decommission phase (Pacca and Horvath, 2002). The study of all the phases is crucial to correctly estimate the overall greenhouse gas emissions.

7.3 Objectives of Life Cycle Assessment

The primary objective of life cycle costing is to evaluate and/or optimize a product's LCCs while satisfying specified performance, safety, reliability, maintainability, and other requirements. The aim is to provide input to decision-making in all phases (especially in the early phases) of a product's life cycle.

7.4 Life Cycle Inventory Analysis

Life cycle inventory analysis involves quantifying the different flows from and to nature for a product system. Inventory analysis involves determining the quantitative values of the materials and the energy inputs and outputs of all process stages within the life cycle. This includes

1. raw materials/energy needs;
2. manufacturing processes;
3. transportation, storage, and distribution requirements;
4. use and reuse; and
5. recycle and end-of-life scenarios such as incineration and land filling.

Development of inventory is done by using input and output data to construct a flow model. A simple schematic of the inputs and outputs from a single stage in a product life cycle system is shown as in Fig. 7.1. This forms the basic building block of life cycle assessments. Further the connected multiple stages in the life cycle of a product are shown as Fig. 7.2.

In case of conventional weir, more concrete is used as compared to rubber dam. Concrete is formed by combination of cement, fine aggregates, and course aggregates. The production of cement requires lot of energy and releases large amount CO_2 emissions. To reduce the

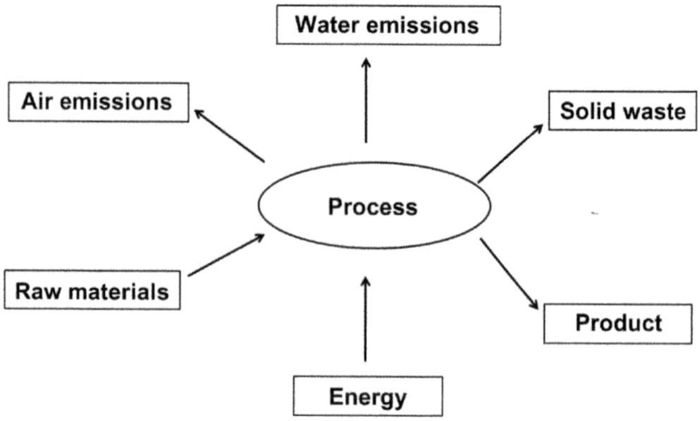

Figure 7.1 Simple schematic of a single stage in a product (Striebig et al., 2015).

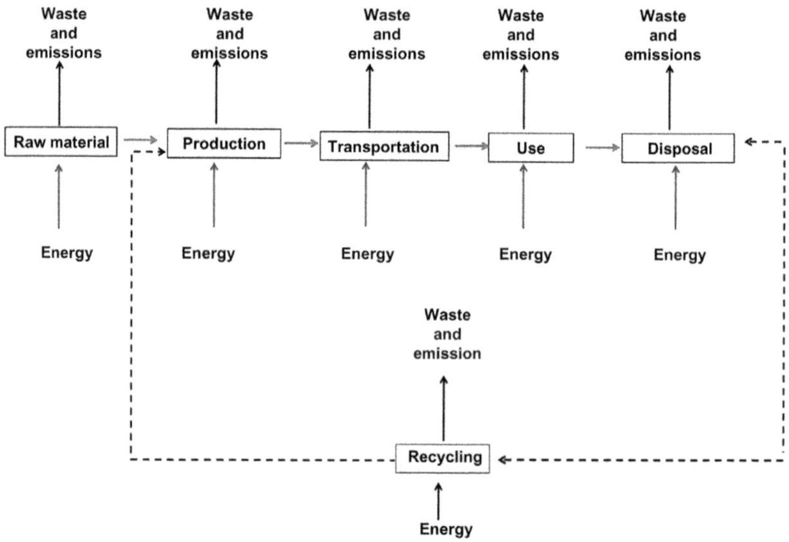

Figure 7.2 A simple schematic of the life cycle of a product (Striebig et al., 2015).

potential emissions during concrete weir, rubber dam is an important alternative. The LCA of concrete weir construction is shown in Fig. 7.3.

7.5 Life Cycle Cost

The LCC is defined as "covering all the costs from project conception to final scrapping and disposal and includes all costs of operation, repairs, maintenance, energy consumption, rentals, insurance, etc., in addition to the initial costs of development and/or acquisition, all

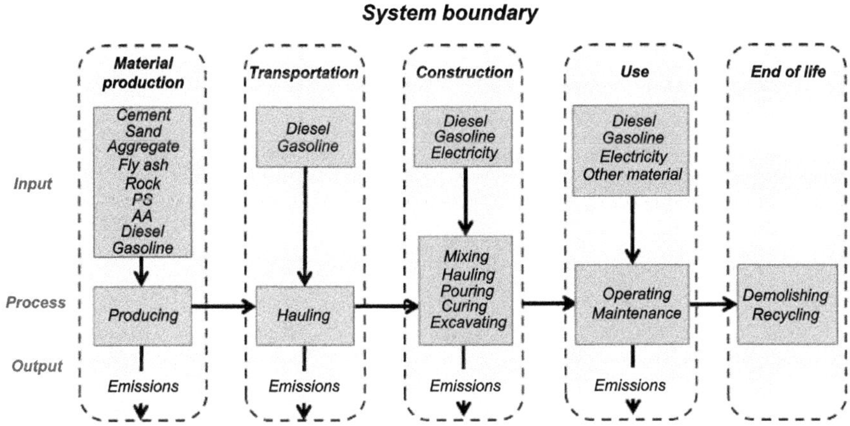

Figure 7.3 The life cycle assessment of concrete weir construction showing emissions during different phases (Liu et al., 2013).

discounted to the same point in time" (Bureau of Indian Standards, 2004). In life cycle costing, comparative LCC assessment of competing alternatives is done over the entire life of the project.

7.6 Case Study

The design discharge and length of weir are important considerations while designing of weir. The water is diverted from the river by using different types of weirs depending upon the feasibility and amount of water to be withdrawn. The different types of weirs to divert water include bush and boulder, raised gravity weir, trench weir, rubber dam, and "mathu bund." The cross-sections of different types of weirs are shown in Fig. 7.4.

The LCC of the above types of weirs used to divert water in hilly area on natural stream is compared. The present worth (PW) method is adopted to calculate LCC (Bureau of Indian Standards, 1991, 1994). All the costs including capital cost, operation and maintenance cost, losses due to head loss, etc. are estimated and the PW is calculated for each cost. The PW is calculated by multiplying expenditure at different periods of time with present worth factors (pwf).

$$\text{Present worth factors (pwf)} = (1 + 0.01 \times i)^{-t}$$

where i is the discount rate per annum and t is the number of years.

Rate of escalation for maintenance and operating costs @ 4% per year. The effect of inflation on LCC calculations will increase the PW

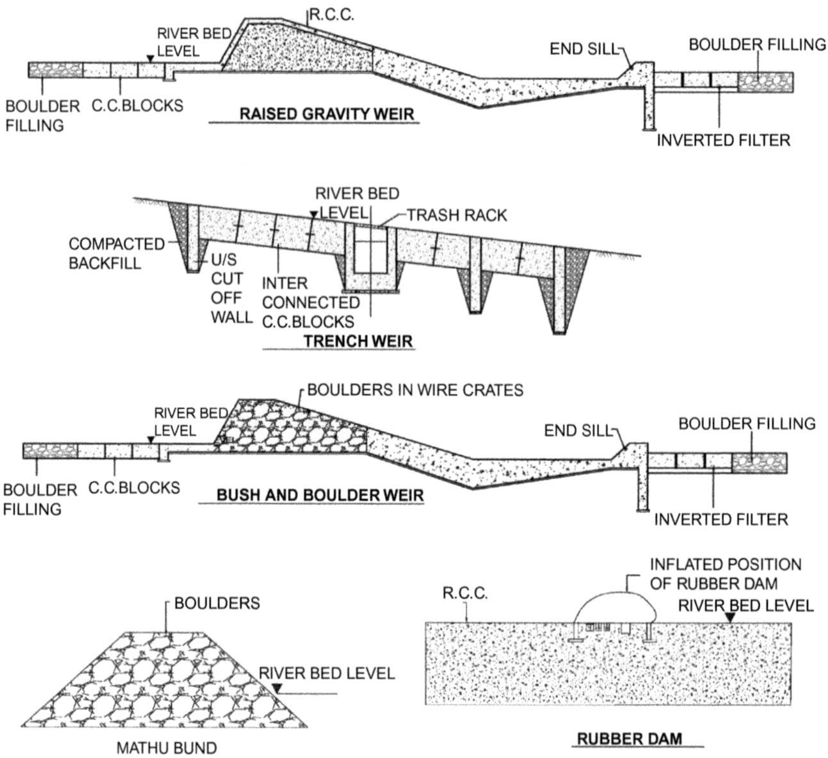

Figure 7.4 Different types of weirs (Islam and Kumar, 2015).

of future expenditure. The expression for pwf and uniform worth factors (upwf), by taking into consideration both the discount rate (i) and escalation rate (r) for a given time period (t), may be obtained as follows:

$$\text{pwf} = \text{b}^{\text{t}}$$

$$\text{upwf} = \frac{b(1 - b^t)}{1 - \text{b}}$$

where

$$b = \frac{(1 + 0.01r)}{(1 + 0.01i)}$$

For $i = 12\%$, $r = 4\%$, and $t = 20$ years, pwf = 0.22714 and upwf = 10.047. The PW of amount spent every year is equal to product of amount and upwf. LCC is the sum of (1) total initial cost; (2) salvage and single expenditure; (3) PW of revenue loss due to head loss, water loss due to flushing, sediment, and repair works; and (4) PW of annual

Table 7.1 Different Projects and Discharge Values for Finding Out Life Cycle Cost (Islam and Kumar, 2015, 2016; AHEC, 2005, 2004)

S. No.	Project Name	Length of Weir (m)	Design Flood Discharge (m^3/s)	Discharge for Hydropower Generation (m^3/s)
1.	Sasoma (J&K)	18	286	1.5, 2.5, 3.5, and 4.5
2.	Umbulung (J&K)	35	382.4	5, 10, 15, and 20
3.	Zunkur (J&K)	33	175	2.5, 5, 7.5, and 10
4.	Pahalgam (J&K)	30.5	425	28.9
5.	Jhanjavati (AP)	60	1800	100

cost (O&M). Different projects and discharge values for finding out LCC are given in Table 7.1.

7.7 Life Cycle Assessment Comparison with Conventional Weirs

The annual cost and LCC of each type of weir is worked out from engineering designs and drawings and inflated weir is compared with other conventional weirs. These sites were selected because of the availability of data regarding weir and site details (Islam and Kumar, 2015, 2016; AHEC, 2005, 2004). Four different discharge values were selected from flow duration curve of each project. Conventional types of weirs were compared with rubber dam constructed with indigenous material manufactured in India as well as imported from international manufacturer. Design of rubber dams and conventional weir was worked out for each of the site.

The initial installation cost was worked out with the help of drawings for each of the project for different type of weirs and is given in Table 7.2.

The operational and maintenance cost for different conventional weirs, Indian rubber dam, and imported rubber dam was worked out by considering cost of repair due to flood, sediment removal cost, cost of operation, and repair works for a typical 2.5 m^3/s of water withdrawal for Sasoma site is given at Table 7.3. In trench weir the head is reduced because the water is withdrawn below the bed level of river. Thus, generation loss due to reduced head (loss) is calculated by taking head loss

Table 7.2 **Initial Installation Cost of Different Types of Weir for Five Different Projects (Islam and Kumar, 2015, 2016; AHEC, 2005, 2004)**

Project	Raised Gravity	Trench Weir	Bush and Bolder	*Mathu* Bund	Indian Rubber Dam	Imported Rubber Dam
Sasoma	120	118	33	28	30	115
Zunkur	227	117	54.32	60	43	180
Umbulung	270	243	252	72	68.4	59.4
Pahalgam	203.3	199.9	55.9	47.4	50.8	194.9
Jhanjavati	499.0	531.0	148.5	126.0	135.0	535.0

Table 7.3 **Operational and Maintenance Cost (in INR $\times$ 10^5) for Different Weirs for a Typical 2.5 m^3/s of Water Withdrawal**

	Types of Weir					
Aspects	Raised Gravity Weir	Trench Weir	Bush and Boulder Weir	*Mathu* Bund	Indian Rubber Dam	Imported Rubber Dam
Major flood damage	0.5	1	3	4.7	0.3	0.2
Sediment removal	0.4	0.6	2	1	0.1	0.4
Cost of operation	1.2	0.5	1.2	1	0.5	1
Minor damages	0.75	0.3	2	0.3	0.2	0.25
Repair works	0.8	1.5	2	3	0.4	0.2
Total (INR)	3.65	3.9	10.2	10	1.5	2.05

as 1.5 m for different water withdrawal values and 3 m in case of Jhanjavati project.

Power loss is calculated by using power equation and total generation loss is calculated by product of power loss and time for which discharge was not available.

$$P = 9.81 \times Q \times H \times \eta$$

Overall efficiency is assumed as 60% and electricity unit cost (kW h) is taken as INR 4. As water is withdrawn/diverted from below the bed level of river in case of trench weir, the water withdrawn contains more sediments as compared with other conventional weirs. For estimation of loss due to 20% of discharge for flushing of sediment, time is taken as 3 months; as for other period, we are having sufficient discharge available. Net head is taken constant for all calculations ($H = 50$ m). The damage due to flood is maximum in case of mathu bund, and repair is necessary.

During repair work, there is loss of discharge and hence loss in power generation. In case of every type of weir, there are some losses due to repair work but in case of mathu bund and bush and boulder type dam is of greater extent hence more loss as compared with other weirs. For calculation of generation loss due to repair work, extra time is taken as 20 and 10 days for "mathu bund" and bush and boulder type, respectively. The revenue loss due to head loss, flushing, and repair works for different sites is shown in Tables 7.4 and 7.5.

The life cycle costing has been worked for a life span of 20 years, a period for which reasonable prediction could be made. Salvage/scrap value has been assumed based on the utility of the product at the end of 20 years. In case of rubber dam, rubber can be used somewhere else after the project and cost of it will add to salvage value. The discount rate is assumed as 12%.

PW of different losses is worked out by assuming inflation of 4% and accordingly values are obtained by multiplying estimated cost with the present value factor. The sum of all these values for 20 years will give net present worth (NPW). Calculation of net present value of different revenue losses for different sites is shown in Table 7.6. The economic life and other basic data for working out the LCC for single discharge of different weirs are given in the Table 7.7.

The result of different losses in terms of PW and NPW is given in Table 7.8.

LCC ($INR \times 10^5$) using PW method of different types of weir for water withdrawal value of 100 m^3/s in case of Jhanjhavati project is given in Table 7.9.

7.7.1 Results

The procedure shown above for working out the LCC was repeated for other four projects. Present net worth of the projects for different discharge values is given in Tables 7.10 and 7.11. The equations which best fit the data were obtained for each type of weir. LCC equations obtained are shown in Table 7.12.

Table 7.4 Revenue Loss due to Head Loss and Power Loss due to Water Loss for Flushing

		Trench Weir					
		Head Loss			Water for Sediment Flushing		
Project	Discharge (m^3/s)	Head (m)	Power Loss (kW)	Cost (INR $\times 10^5$)	Head (m)	Power Loss (kW)	Cost (INR $\times 10^5$)
Sasoma	2.5	1.5	22.07	7.72	50	147.15	12.7
Zunkur	5	1.5	44.14	15.468	50	294.3	25.42
Umbulung	10	1.5	88.29	30.94	50	588.6	50.85
Pahalgam	28.9	1.5	255.15	894	50	1701.05	146.97
Jhanjavati	100	3	1765.8	618.4	50	5886	169.5

Table 7.5 Revenue Loss due to Water Loss due to Repair

	Mathu Bund			B&B Weir
Project	Head (m)	Power Loss (kW)	Cost (INR × 10^5)	Cost (INR × 10^5)
Sasoma	50	735.75	14.13	7.06
Zunker	50	1471.5	28.26	14.13
Umbulung	50	2943	56.52	28.26
Pahalgam	50	8505.27	163.3	81.65
Jhanjavati	50	29,430	565	282.52

Annual cost of rubber dam is also less as compared with other conventional weirs (Islam and Kumar, 2016). The equations for calculating the annual cost are given in Table 7.13.

These results show that annual cost of rubber dam is less as compared to other conventional type weir. Equations show that annual cost of Indian rubber dam is lowest while that of trench weir is more.

7.7.2 Reference Standards

As a part of ISO 14000 series of standards, ISO in 2000 has come out with the following standards:

1. ISO 14040: Environmental Management—LCA—Principles and Framework.
2. ISO 14041: Environmental Management—LCA—Inventory Analysis.
3. ISO 14042: Environmental Management—LCA—Impact Assessment.
4. ISO 14043: Environmental Management—LCA—Interpretation

There are some other international standards for LCA and are as follows:

1. ISO 9000:2005, Quality management systems—Fundamentals and vocabulary
2. ISO 14001:2004, Environmental management systems—Requirements with guidance for use

Table 7.6 Net Present Worth (NPW) (INR $\times 10^5$) of Different Revenue Losses due to Head Loss of Trench Weir for Sasoma Project

Year	0	1	2	3	4	5	6	7	8	9	NPW
Estimated cost	7.72	8.03	8.35	8.68	9.03	9.39	9.77	10.16	10.57	10.99	**83.53**
Present worth	7.72	7.17	6.66	6.18	5.74	5.33	4.95	4.60	4.27	3.96	
Year	10	11	12	13	14	15	16	17	18	19	
Estimated cost	11.43	11.88	12.36	12.85	13.37	13.9	14.46	15.04	15.64	16.26	
Present worth	3.68	3.42	3.17	2.95	2.74	2.54	2.36	2.19	2.03	1.89	

NPW, Net present worth.

Table 7.7 Basic Data and Assumptions of Five Projects for Single Discharge for Working out Life Cycle Cost of Different Weirs

Project	Item	Raised Gravity	Trench Weir	Bush and Bolder	Mathu Bund	Indian Rubber Dam	Imported Rubber Dam
Sasoma SHP	Economical life (years)	20	20	10	5	20	20
	Discount rate (%)	12	12	12	12	12	12
	Total initial cost (INR × 10^5)	120	118	33	28	30	115
	Operational and maintenance cost (annual) (INR × 10^5)	3.65	3.9	10.2	10	1.5	2.05
	Single expenditure at the end of 10 years (INR × 10^5)	29	42	19.8	61.6	18	69
	Salvage value after 20 years (INR)					9	34.5
Jhanjavati	Total initial cost (INR × 10^5)	499.0	531.0	148.5	126.0	135.0	535.0
	Operational and maintenance cost (annual) (INR × 10^5)	15.2	17.6	45.9	45.0	6.8	9.5
	Single expenditure at the end of 10 years (INR × 10^5)	120.6	189.0	89.1	277.2	81.0	321.0
	Salvage value after 20 years (INR)					40.5	160.5
Pahalgam SHP	Total initial cost (INR × 10^5)	203.3	199.9	55.9	47.4	50.8	194.9
	Operational and maintenance cost (annual) (INR × 10^5)	6.2	6.6	17.3	16.9	2.5	3.5
	Single expenditure at the end of 10 years (INR × 10^5)	49.1	71.2	33.6	104.4	30.5	116.9
	Salvage value after 20 years (INR × 10^5)					15.3	58.5

Table 7.7 **Basic Data and Assumptions of Five Projects for Single Discharge for Working out Life Cycle Cost of Different Weirs—cont'd**

Project	Item	Raised Gravity	Trench Weir	Bush and Bolder	Mathu Bund	Indian Rubber Dam	Imported Rubber Dam
Umbulung SHP	Total initial cost (INR × 10^5)	227	117	54.32	60	43	180
	Operational and maintenance cost (annual) (INR × 10^5)	6.9	3.9	16.8	21.4	2.2	3.2
	Single expenditure at the end of 10 years (INR × 10^5)	54.9	41.6	32.6	132.0	25.8	108.0
	Salvage value after 20 years (INR)					13	54
Zunkur SHP	Total initial cost (INR × 10^5)	270	243	252	72	68.4	59.4
	Operational and maintenance cost (annual) (INR × 10^5)	8.21	8.03	77.89	25.71	3.42	1.06
	Single expenditure at the end of 10 years (INR × 10^5)	65.25	86.49	151.20	158.40	41.04	35.64
	Salvage value after 20 years (INR)					20	18

Table 7.8 NPW (INR $\times 10^5$) of Different Revenue Losses

	Trench Weir				Bush and Bolder		*Mathu* Bund	
	Loss due to Head Loss		Loss due to Flushing of Water		Loss due to Repair Work for Flood Damage		Loss due to Repair Work for Flood Damage	
Site	EC	NPW	EC	NPW	EC	NPW	EC	NPW
Sasoma	7.72	83.53	12.70	137.4	7.06	76.39	14.13	152.89
Zunker	15.468	167.36	25.42	274.8	14.13	152.74	28.26	305.49
Umbung	30.94	334.46	50.85	549.6	28.26	305.5	56.52	610.98
Pahalgam	894	9673	146.97	1590.2	81.65	883.45	163.3	1766.9
Jhanjavati	618.4	6691.0	169.50	1833.9	285.53	3089.4	565.0	6113.27

NPW, Net present worth; *EC*, estimated cost.

Table 7.9 Life Cycle Cost (INR × 10^5) Using Present Worth Method of Different Types of Weir for Water Withdrawal Value of 100 m^3/s in Case of Jhanjhavati Project

Item	Raised Gravity		Trench Weir		Bush and Bolder		*Mathu Bund*		Indian Rubber Dam		Imported Rubber Dam	
	EC	PW	EC	PW	EC	PW	EC	PW	EC	PW	EC	PW
Total initial cost	580	580	531	531	148.5	148.5	126	126	135	135	535	535
Salvage and single expenditure												
(a) Expenditure after 5 years					74.3	42.1	277.2	157.3				
Amt × pwf = (pwf = 0.5674)												
(b) Expenditure after 10 years	120.8	38.9	189.6	61.1	326.7	105.2	277.2	89.3	84.4	27.2	321	103.4
Amt × pwf = (pwf = 0.3220)												
(c) Expenditure after 15 years					207.9	38	352.8	64.5				
Amt × pwf = (pwf = 0.1827)												
(d) Salvage after 20 years									40.5	4.2	160.5	16.6
Amt × pwf = (pwf = 0.1037)												
NPW of losses				8525		3089.3		6113.2				
Annual cost (O&M)	15.2	152.5	17.6	176.3	45.9	461.2	45	452.1	6.8	67.8	9.5	95.8
Amt × upwf Upwf = 10.047												
Life cycle cost		**771.4**		**9293.4**		**3884.3**		**7002.3**		**230**		**734.2**

NPW, Net present worth; *EC*, estimated cost; *upwf*, uniform worth factors; *pwf*, present worth factors.

Table 7.10 Life Cycle Cost (INR × 10^5) of Different Types of Weirs for Different Discharge Values

Project	Discharge (m^3/s)	Type of Weir					
		Raised Gravity	Trench Weir	Bush and Bolder	*Mathu Bund*	Indian Rubber Dam	Imported Rubber Dam
Sasoma SHP	1.5	143.0	299.6	199.7	255.9	44.2	141.3
	2.5	166.1	391.6	254.2	340.9	50.0	154.2
	3.5	202.6	492.2	315.1	434.6	57.7	196.0
	4.5	246.3	588.4	404.6	546.5	73.7	226.9
Umbulung SHP	5	515.4	1066.9	752.9	1040.6	130.1	413.4
	10	617.3	1689.1	1076.4	1641.5	155.4	494.9
	15	738.5	2255.2	1414.9	2255.8	185.9	592.8
	20	851.1	2818.1	1721.2	2832.6	203.5	678.7
Zunkur SHP	2.5	357.0	791.7	593.4	933.4	105.7	320.9
	5	607.9	1405.3	1019.2	1585.1	174.4	549.0
	7.5	838.2	1988.9	1414.9	2252.6	240.3	761.1
	10	1158.3	2715.0	1874.1	2986.3	332.0	1055.9
Jhanjhavati project	100	771.4	9293.4	3884.3	7002.3	230.0	734.2
Pahalgam SHP	28.9	281.3	11552.2	1155.9	2085.5	84.6	262.3

Table 7.11 Life Cycle Cost (INR × 10^5) Per Unit Length of Different Types of Weirs for Different Discharge Values in Case of Sasoma Hydroproject

Project	Discharge (m^3/s)	Type of Weir					
		Raised Gravity	Trench Weir	Bush and Bolder	*Mathu Bund*	Indian Rubber Dam	Imported Rubber Dam
Sasoma SHP	1.5	7.94	16.64	11.09	14.22	2.46	7.85
	2.5	9.23	21.76	14.12	18.94	2.78	8.57
	3.5	11.26	27.34	17.51	24.14	3.21	10.89
	4.5	13.68	32.69	22.48	30.36	4.09	12.61
Umbulung SHP	5	14.73	30.48	21.51	29.73	3.72	11.81
	10	17.64	48.26	30.75	46.90	4.44	14.14
	15	21.10	64.43	40.43	64.45	5.31	16.94
	20	24.32	80.52	49.18	80.93	5.81	19.39
Zunkur SHP	2.5	10.82	23.99	17.98	28.28	3.20	9.72
	5	18.42	42.58	30.88	48.03	5.28	16.64
	7.5	25.40	60.27	42.88	68.26	7.28	23.06
	10	35.10	82.27	56.79	90.49	10.06	32.00
Jhanjavati project	100	25.71	309.78	129.48	233.41	7.67	24.47
Pahalgam SHP	28.9	9.22	378.76	37.90	68.38	2.77	8.60

Table 7.12 **Equations for Total Life Cycle Cost (INR × 10^5) Per Unit Length for Different Discharge (Islam and Kumar, 2016)**

Type of Weir	Life Cycle Cost (C)	Coefficient of Determination (R^2)
Trench weir	$\mathbf{10.286e^{0.2954Q}}$	0.8427
Mathu bund	$\mathbf{4.2235e^{0.4161Q}}$	0.8403
Bush and boulder	$\mathbf{8.5431e^{0.2428Q}}$	0.8698
Raised weir	$\mathbf{9.0295e^{0.1142Q}}$	0.8136
Imported rubber dam	$\mathbf{8.7806e^{0.0948Q}}$	0.6757
Indian rubber dam	$\mathbf{3.5979e^{0.0535Q}}$	0.1616

Q is the water withdrawal for hydropower generation and C is the life cycle cost (lakhs). From the results it is concluded that life cycle cost of rubber dam is less than other types of conventional weirs.

Table 7.13 **Equations for Total Annual Cost (INR × 10^5) Per Unit Length for Different Discharge (Islam and Kumar, 2015)**

Type of Weir	Annual Cost (C) (INR × 10^5)	Coefficient of Determination (R^2)
Trench weir	$\mathbf{0.9418e^{0.2452Q}}$	0.9963
Mathu bund	$\mathbf{0.812e^{0.2586Q}}$	0.974
Bush and boulder	$\mathbf{0.6878e^{0.234Q}}$	0.9897
Raised weir	$\mathbf{0.7305e^{0.2001Q}}$	0.9746
Imported rubber dam	$\mathbf{0.7285e^{0.1782Q}}$	0.9467
Indian rubber dam	$\mathbf{0.3877e^{0.1744Q}}$	0.9364

Q is the water withdrawal for hydropower generation and C is the annual cost (lakhs).

3. ISO 14004:2004, Environmental management systems—General guidelines on principles, systems and support techniques
4. ISO 14020, Environmental labels and declarations—General principles
5. ISO 14021, Environmental labels and declarations—Self-declared environmental claims (Type II environmental labeling)
6. ISO 14025, Environmental labels and declarations—Type III environmental declarations—Principles and procedures

7. ISO 14031, Environmental management—Environmental performance evaluation—Guidelines
7. ISO/TR 14032, Environmental management—Examples of environmental performance evaluation (EPE)
9. ISO/TR 14047, Environmental management—Life cycle impact assessment—Examples of application of ISO 14042
10. ISO/TS 14048, Environmental management—Life cycle assessment—Data documentation format
11. ISO/TR 14049, Environmental management—Life cycle assessment—Examples of application of ISO 14041 to goal and scope definition and inventory analysis
12. ISO 14050, Environmental management—Vocabulary
13. ISO/TR 14062, Environmental management—Integrating environmental aspects into product design and development
14. ISO 14063, Environmental management—Environmental communication—Guidelines and examples
15. ISO 14064-1, Greenhouse gases—Part 1: Specification with guidance at the organization level for quantification and reporting of greenhouse gas emissions and removals
16. ISO 14064-2, Greenhouse gases—Part 2: Specification with guidance at the project level for quantification, monitoring and reporting of greenhouse gas emission reductions or removal enhancements
17. ISO 14064-3, Greenhouse gases—Part 3: Specification with guidance for the validation and verification of greenhouse gas assertions
18. ISO Guide 64, Guide for the inclusion of environmental aspects in product standards

7.8 Conclusions

LCA is a useful tool for evaluating the comprehensive impacts from the manufacture, use, and disposal of products or system or service. The result of an LCA can illustrate comparative differences in environmental and health impacts, such as whether product/service A or product/service B has a greater likely impact on climate change.

Table 7.14 Some Currently Available LCA Software

Tool	Website	Developers
ECO-it	www.pre-sustainablity.com	PRé Consultants
eiolca.net	www.eiolca.net	Green Design Institute at Carnegie Mellon University
GaBi	www.gabi-software.com	PE-International
GREET	www.greet.es.anl.gov	Argonne National Laboratory
IDEMAT	http://www.idemat.nl/	Delft University of Technology
SimaPro	www.pre-sustainability.com	PRé Consultants
SULCA	www.vtt.fi	VTT Technical Research Centre of Finland
Umberto	www.ifu.com/en/umberto/	ifu Hamburg GmbH

From the study, following conclusions are made:

1. LCC of imported rubber dam as well as Indian rubber dam is found to be less than all the conventional types of weirs. Average LCC of imported rubber dam is about three times more than Indian rubber dam.
2. With the comparative analysis, it is found that average LCC of raised gravity weir, bush and boulder, *mathu bund*, and trench weir is 2, 3, 5, and 6 times, respectively, more than imported rubber dam.
3. Inflatable weirs are easy to inflate and deflate resulting in passage of sediments and boulders easily from upstream to downstream of rubber dam due to which damage due to flood is reduced. This reduces the cost of operation and maintenance, repair works, sediment removal, and flood damage. Rubber dams are simple, reliable, and flexible structure resulting in wide range of application in hydropower, irrigation, and flood control and water resource management. Because of low LCC, its application should be recommended.
4. Due to the large quantities of data that have to be collected and processed for any LCA, computer software is usually used to perform most LCAs. Table 7.14 shows examples of some currently available LCA software and their developers.

References

AHEC, Jan 2005. DPR of Sasoma Hydro Project Prepared by AHEC, IIT Roorkee, Roorkee, India. AHEC/PDC, Sasoma.

AHEC, Jan 2004. DPR of Umbulung Pashkum and Zunkur hydroelectric projects prepared by Kargil Renewable Energy Development Agency, District Kargil, (J&K). AHEC/PDC, Pashkum, India.

Bureau of Indian Standards, 1991, 1994. Indian standard life cycle costing. In: IS 13174[Part 1(1991) and 2(1994)], New Delhi.

Bureau of Indian Standards, 2004. Indian standard dependability management application guide—life cycle costing. In: IS 15474:2004, New Delhi.

Emergency Management Ontario (EMO), 2005. The Ontario Provincial Hazard Identification and Risk Assessment. Ministry of Community Safety and Correctional Services.

Islam, Su, Kumar, A., 2016. Inflatable dams for SHP projects. Renew. Sustain. Energy Rev. 57, 945–952.

Islam, S.U., Kumar, A., 2015. Evaluation of rubber dams for SHP in India. Indian J. Sci. Technol. 8 (28), 755.

Jackson, S., Sleigh, A., 2000. Resettlement for China's Three Gorges Dam: socio-economic impact and institutional tensions. Commun. Post-Commun. Stud. 33.2, 223–241.

Jensen, A.A., 1997. Life Cycle Assessment. A Guide to Approaches, Experiences and Information Sources, 121. European Environment Agency.

Liu, C., et al., 2013. Life-cycle assessment of concrete dam construction: comparison of environmental impact of rock-filled and conventional concrete. J. Construct. Eng. Manage. 139.12, A4013009.

Pacca, S., Horvath, A., 2002. Greenhouse gas emissions from building and operating electric power plants in the Upper Colorado River Basin. Environ. Sci. Technol. 36, 3194–3200.

Rickard, C., Day, R., Purseglove, J., 2003. River Weirs–Good Practice Guide Guide-Section A. Unpublished report for Environment Agency.

Stewart, R.A., 2000. Dam risk management. In: ISRM International Symposium. International Society for Rock Mechanics.

Striebig, B., Ogundipe, A.A., Papadakis, M., 2015. Engineering Applications in Sustainable Design and Development, SI ed. Cengage Learning, p. 755.

Index

Note: Page numbers followed by "*f*" and "*t*" refer to figures and tables, respectively.

A

Additives, 26–27
Adhesion
  adhesive systems, 27–28
  fabrics, 40–41
  of textiles with rubber, 27–28
Aeration, 20
Aerators for improving water quality, 20–22
Air bubbles, 43
Air retention, 28–29
Air rubber dam, 49
Air-filled rubber dam, 5–9, 6*f*, 48, 50*f*
Air-inflated dams, 68–69
Airtight seal, 62
American Society for Testing and Materials (ASTM), 93–94
American Society of Mechanical Engineers (ASME), 93
American Welding Society (AWS), 94
Anchor bolts, 5–9, 59, 62
  fixed on foundation, 60*f*
  on walls, 59, 61*f*
Anchor plates, 54–55
Anchoring connections, 89
Anchoring method/system, 51–52, 56–58
  anchorage
    of dam bag with concrete base, 56–57
    of rubber dam sheets with side walls of abutment or piers, 57–58
  double anchored rubber dam, 52, 52*f*
  single anchored rubber dam, 51, 52*f*
  types, 58–59
Angle of inclination, 5–9
Aquatic ecosystem, 22
Aramid, 33–34
Artificial recharge of groundwater from rubber dams, 19
ASME. *See* American Society of Mechanical Engineers (ASME)
Assembly of components, 43
ASTM. *See* American Society for Testing and Materials (ASTM)
Autoclave vulcanization method, 43–44
Automatic control mechanism of rubber dam, 92*f*
AWS. *See* American Welding Society (AWS)

B

Biodiversity, rubber dam impact on, 23
Biological oxygen demand (BOD), 13–14
Bladder
  connections, 89
  rubber, 5–9, 51
  rubber dam bladder anchorage, 86
  rubber dam bladder specification, 86

Blue-green algae, 22
BOD. *See* Biological oxygen demand (BOD)
Bonding of dam body, 53–56
  considerations for construction of rubber dam body, 55
  durability of rubber dam, 56
  physical property of rubber dam, 54–55
  specification of rubber dam, 54
Bookend rubber dam, 5–9, 7*f*
Bucket type of deflation systems, 74–75
Bulk defects, 44, 45*f*
Bush and boulder weir, 101–102, 105, 106*f*
  data and assumptions of projects, 113*t*
  initial installation cost, 107, 108*t*
  LCC, 116*t*, 117*t*, 119*t*
  NPW, 115*t*
  operational and maintenance cost, 107–108, 108*t*
  revenue loss, 111*t*
  total annual cost, 119*t*
Butyl rubber (IIR), 28–29, 31–32

C

Calendering operations, 40–42
  on rubber compound, 37–38
Calenders, 41–42
Canvas, 5
Caprolactam, 33–34
Central Institute for Research on Cotton Technology, 82
Chandeshwar-I rubber dam Kurda Orissa, India, 91, 91*f*
Check dams, 5–9
Chloroprene rubber (CR), 26–27, 31–32, 53, 56
Chlorosulfonated polyethylene rubber (CSM), 28–29, 31–32
cis-BR, 31–32
Clamp plate fixed with rubber sheet, 59, 61*f*
Clamping
  anchor bolts, 62
  plates, 5–9, 54–55, 62
  by punching fabric, 62
  without punching fabric, 62
  rubber fabric with concrete, 59–62
  system, 5–9, 58–59
Cleaning foundation anchor bolts, 59, 60*f*
Coanda effect, 75
Coated fabric, 41–42
Coating, 41–42, 42*f*
Coefficient of discharge, 78–80
Cold SBR, 31–32
Concrete, 34–36, 103–104
  concrete-based weir impacts, 100
  wedge blocks, 58
Consumers, 22
Continuous vulcanization method, 43–44
Control shaft, 88
Conventional weirs, 101
  LCA comparison with, 107–120
Conveyance losses, 15, 15*t*
Cotton, 33–34
CR. *See* Chloroprene rubber (CR)
CSM. *See* Chlorosulfonated polyethylene rubber (CSM)
Customization, 45–46

D

Dams, 2–3, 3*f*
  on basis of structure, use, and material, 3*f*
  body, 48–49, 52
  functions, 4*f*
Deflation of rubber dam, 5–9, 67–68
  design criteria, 84–85
  emergency mechanical deflation system, 85
  maintenance during, 81–83
  phenomenon, 74–77, 83*f*
Deflected rubber membrane, 77–78
Deflectors, 54–55

Design flow event with rubber dam bladder deflated, 85
Diatoms (Bacillariophyceae), 22
DIN. *See* German Institute for Standardization (DIN)
Dip coating, 40–41
Dispersion, 38–40
Dissolved oxygen (DO), 13–14
Distribution, 38–40
Diurnal storage headwork, 99–100
Diversion headworks, 99–100
DO. *See* Dissolved oxygen (DO)
Double anchored rubber dam, 52, 52*f*
Double anchoring line, 54, 81, 81*f*
Double clamp systems, 5–9
Double-line anchoring, 56–57
Downstream hydrostatic force, 55
Drainage shaft, 88–89
Durability of rubber dam, 56

E

Earthquake force, 63
Ecological manifestation, 22–23
  animals, 23
  consumers, 22
  producers, 22
Electric-operated aerators, 20, 21*f*
Electrical type of deflation systems, 74–75
Electricity, 43–44
Elongation, 34
Emergency mechanical deflation system, 85
Environment, rubber dam impact on, 23
Ethylene propylene diene monomer (EPDM), 28–29, 31–32, 54–56
Eutrophic zone, 22
Evaporation losses from pond, 16

F

Fabrics adhesion, 40–41
Fabridams. *See* Rubber dams
Fatigue resistance, 34
Fillers, 31–32
Filling
  medium, 5–9
  shaft, 88
  time, 15
Fin, 54–55
Fitment clamps and anchors, 5–9
Fixing
  blocks, 59, 60*f*
  of rubber sheet, 59
Flexibility, 34
Float type of deflation systems, 74–75
Floating aerators, 20
Flooding debris and sediment, 82–83
Fluid adoption for rubber dam, 20
Four-roll calender, 41–42
Freshwater, 12
Friction ratio, 40
Frictioning, 41–42, 42*f*

G

Galvanized steel, 54
Gauge, 34
German Institute for Standardization (DIN), 94
Glen Canyon dam
  CO_2 emission during construction of, 100
  decommissioning phase, 100
  watershed management, 100–101
Government Water Policies, 11
Green algae, 22
Groundwater
  augmentation, 13
  conservation of groundwater resources, 12
  levels, 12

H

Height determination of dam, 58
Herco Rub Technical Article, 30*t*
Hot air, 43–44

Hot emulsion SBR, 31–32
Hsieh model, 78–80
Hydraulic structures, 4–5
Hydrodynamic forces, 55, 63
Hydrostatic force, 63

I

IFMDs. *See* Inflatable flexible membrane dams (IFMDs)
Inclined dam, 56
Inclined rubber dam, 50–51, 51*f*
Incorporation, 38–40
Indian Rubber Manufactures Research Association, 82
Inertia force, 63
Inexpensive operating system, 4–5
Inflatable dam, 101
Inflatable flexible membrane dams (IFMDs), 67–68
Inflatable rubber dams, 4–5, 6*f*, 48–49
Inflatable weirs. *See* Rubber dams
Inflation of rubber dam, 5–9, 68–69
  design criteria, 84
  maintenance during, 81–83
  medium of inflation, 70–71
  optimization of suitable inflation deflation mechanism, 91
  overflow of inflated rubber dams, 78
  phenomenon, 74–77, 83*f*
  time, 90–91
Internal mill, 38–40, 39*f*
Internal mixer, 38–40
Internal water/air pressure, 55
International Organization for Standardization (ISO), 94–95
ISO. *See* International Organization for Standardization (ISO)
Isoprene, 29–31

J

Jhanjavati project, 113*t*
  discharge values, 107*t*
  initial installation cost, 107, 108*t*
  LCC, 117*t*
  NPW, 115*t*
  revenue loss, 110*t*, 111*t*

L

Land Development Technical Standard of Rubberized Fabric Inflatable Weirs (LDTRC), 48
Latex, 40–41
LCA. *See* Life cycle assessment (LCA)
LCC. *See* Life cycle cost (LCC)
LDTRC. *See* Land Development Technical Standard of Rubberized Fabric Inflatable Weirs (LDTRC)
Leno weave, 35*t*
Life cycle assessment (LCA), 102–103
  case study, 105–107
  comparison with conventional weirs, 107–120
    power loss, 110*t*
    reference standards, 111–120
    results, 109–111
    revenue loss, 110*t*, 111*t*
  of concrete weir construction, 105*f*
  life cycle inventory analysis, 103–104
    life cycle of product, 104*f*
    single stage in product, 104*f*
  objectives, 103
  software, 121*t*
Life cycle cost (LCC), 101–102, 104–105
Linyi City of Shandong State in China, 16
Low-temperature crystallization, 31–32

M

Manufacturing process, 37
  layout, 38*f*

mixing sequence for preparation of rubber compound, 39*f*
of reinforced rubber sheet for rubber dam, 37, 38*f*
Mastication, 38–40
"Mathu bund", 101–102, 105, 106*f*
initial installation cost, 107, 108*t*
LCC, 116*t*, 117*t*, 119*t*
NPW, 115*t*
operational and maintenance cost, 107–108, 108*t*
revenue loss, 111*t*
total annual cost, 119*t*
Matt weave, 35*t*
Medium
of filling material, 49
of inflation, 70–71
Middle layer of rubber sheets, 55
Mixing rubbers, 38–40
Model studies, 14
Moisture in textiles, 41–42
Musi River in City of Hyderabad, India, 19

N
Nakdonggang River in Daegu City of South Korea, 16–19
Nappe adherence instability, 75
Natural rubber (NR), 5–9, 26–27, 31–32
Net present worth (NPW), 109, 112*t*
Nylon, 33–34
Nylon 6, 33–34
Nylon 66, 33–34

O
O&M cost. *See* Operational and maintenance cost (O&M cost)
Oil, 43–44
Open mills, 38–40, 40*f*
Operational and maintenance cost (O&M cost), 16, 107–108, 108*t*
Operational equipment of rubber dam, 87–89
Outer layer of rubber sheets, 55
Overflow above and deflated flexible membrane dam, 77–78
anchor details for rubber dam installation, 82*f*
coefficient of discharge, 78–80
fluid-structure instabilities, 80*t*
maintenance during inflation and deflation, 81–83
observations and recommendations to prevent vibration cases, 79*t*
overflow of inflated rubber dams, 78
single anchor line and double anchor line, 81*f*

P
Pahalgam SHP project, 113*t*
discharge values, 107*t*
initial installation cost, 107, 108*t*
LCC, 117*t*
NPW, 115*t*
revenue loss, 110*t*, 111*t*
PBR. *See* Polybutadiene rubber (PBR)
Percolation test, 63–64
Periphyton, 22
Phytoplankton, 22
Pillow-like rubber dam, 49–50, 51*f*, 56–57
Pipe connections, 89
Plain weave, 35*t*
Plasticizers, 31–32
Polybutadiene rubber (PBR), 28–29
Polyester, 33–34
Polyisoprene rubber, 29–31
Polymer composites, 4–5
Power loss calculation, 108
Present worth factors (pwf), 105–106
Present worth method (PW method), 105
Press vulcanization method, 43–44
Producers, 22
Pump shaft, 87

Punching rubber sheet for fixing, 59, 61*f*
PW method. *See* Present worth method (PW method)
pwf. *See* Present worth factors (pwf)

Q

Quality
- check operations for defects, 45–46
- test for reinforced rubber sheet for rubber dam, 44, 44*f*

R

Rainwater, 14
- harvesting, 13

Raised gravity weir, 101–102, 105, 106*f*
- data and assumptions of projects, 113*t*
- initial installation cost, 107, 108*t*
- LCC, 116*t*, 117*t*, 119*t*
- operational and maintenance cost, 107–108, 108*t*
- total annual cost, 119*t*

Raw materials
- advantages, 28*f*
- construction in fabric structure, 35*t*
- for preparation of uncured rubber compound, 27*f*

Rayon, 33–34
RDSS. *See* Rubber dam system supplier/manufacturer (RDSS)
Reinforced concrete, 34–36
Reinforced rubber sheet, 26–27, 41–42
- assembly, 43, 43*f*
- manufacturing process, 37, 38*f*

Resorcinol–formaldehyde–latex bonding system (RFL bonding system), 40–41
River morphology, 14
Rubber dam system supplier/manufacturer (RDSS), 84, 86
Rubber dams, 4–5, 26–27, 48–53, 57, 67–68, 68*f*, 101–102, 105, 106*f*
- advantages, 71–72
- anchoring method, 51–52
- automatic control mechanism of rubber dam, 92*f*
- bladders, 55
- cross-section, 70*f*
- data and assumptions of projects, 113*t*
- design criteria for, 83–89
- fluid adopted for, 20
- forces
  - acting on dam, 76*f*
  - acting on element, 76*f*
- foundation for, 62–64
  - construction procedure for lying of foundation concrete, 63–64
  - loads considered for design of concrete base, 63
  - selection of site for suitability of foundation, 63
- in India, 18*t*, 19*f*
- inflation and deflation phenomenon, 74–77
- inflation time, 90–91
- initial installation cost, 107, 108*t*
- installations, 16–19, 72–74
- LCC, 116*t*, 117*t*, 119*t*
- medium of filling material, 49
- medium of inflation, 70–71
- operational and maintenance cost, 107–108, 108*t*
- operational steps for water-filled dam, 89–90, 90*f*
- optimization of suitable inflation deflation mechanism, 91
- overflow, 75*f*
- parts, 69–70
- reference standards, 93–95
- safe and user-friendly operation, 91–92
- seam of rubber bag, 52–53

sediment discharge system of rubber dam, 93*f*
seismic national and international requirements, 89
selection of type, 68–69
side slope, 49–51
special types, 64–65
storage, 15–16
capacity and filling time, 15
evaporation losses from pond, 16
seepage losses from pond, 15
total annual cost, 119*t*
for water conservation and management, 13–14
Rubber(s), 25–27
bladder, 5–9, 51
body, 5
chemical structure, 31*f*
compounder, 36
compounding, 36
compounds, 40
formulation, 26–27
molecules, 38–40
properties, 30*t*
rubber/textile composites, 27–28
sheets, 55
Rubberized fabrics, 82
membrane, 53

S

Safe and user-friendly operation, 91–92
Safe bearing capacity, test for, 63–64
Santa Ana River Inflatable Rubber Dam Project, 19
Sasoma SHP project, 107–108, 113*t*
discharge values, 107*t*
initial installation cost, 107, 108*t*
LCC, 117*t*
NPW, 112*t*, 115*t*
revenue loss, 110*t*, 111*t*
SBR. *See* Styrene butadiene rubber (SBR)
SCADA system. *See* Supervisory control and data acquisition system (SCADA system)
Seam of rubber bag, 52–53
seamed rubber dam, 53
seamless rubber dam, 53
Seamed rubber dam, 53
Seamless rubber dam, 53
Sediment discharge system of rubber dam, 93*f*
Seepage losses from pond, 15
Seismic national and international requirements, 89
Self load of dam in empty and full condition, 63
Shield-type rubber dam, 5–9, 8*f*, 64–65
Side slope, 49–51
Silt or ice load, 63
Simple operating system, 4–5
Single anchor line, 81, 81*f*
Single anchored rubber dam, 51, 52*f*
Single-line anchoring, 56
Slope of abutment or side piers, 58
Smooth functioning of rubber dam, 5–9
Society, rubber dam impact on, 23
Soda cellulose, 33–34
Solar-operated aerators, 20, 21*f*
Solid woven fabric structures, 35*t*
Solution BR, 31–32
Solution SBR, 31–32
Solvents, 43
Spread coating, 41–42
Stainless steel, 54
Steam, 43–44
Steel clamps with anchor bolts, 58–59
Steel shield plate, 65
Strain-induced crystallization, 31–32
Stress warp fabric structures, 35*t*
Stretching unit, 40–41, 41*f*
Styrene butadiene rubber (SBR), 26–27
latex–Resorcinol-treated fabric, 82
solution, 31–32

Supervisory control and data acquisition system (SCADA system), 72–74
Surface defects, 44, 45*f*
Surface water resources, conservation of, 12
Synthetic
  fibers, 34
  materials, 56
  rubbers, 5–9

T

Taoyuan, river Yihe, Linyi, 18*t*
Tear strength, 34
Textile fibers, 40–41
  general physical properties of, 33–34, 34*t*
Textile/rubber composites, 34
Three-roll calender, 41–42
Transverse length of membrane, 56
Trench weir (Tyrolean), 101–102, 105, 106*f*
  data and assumptions of projects, 113*t*
  initial installation cost, 107, 108*t*
  LCC, 116*t*, 117*t*, 119*t*
  NPW, 115*t*
  operational and maintenance cost, 107–108, 108*t*
  revenue loss, 110*t*
  total annual cost, 119*t*
Triaxial woven fabric structures, 35*t*
Twill, 35*t*
Two-roll calender, 41–42
Two-roll mill, 40

U

Ultraviolet (UV), 5
Umbulung SHP project, 113*t*
  discharge values, 107*t*
  initial installation cost, 107, 108*t*
  LCC, 117*t*
  NPW, 115*t*
  revenue loss, 110*t*, 111*t*
Uniform present worth factors (upwf), 105–106
Uplift force, 63
Upstream hydrostatic force, 55
upwf. *See* Uniform present worth factors (upwf)
Usable water, 11–12
Utilizable water resources, 14
UV. *See* Ultraviolet (UV)

V

Vertical side slopes, 49–50
Vinyl Pyridine latex–Resorcinol-treated nylon fabric, 82
Viscosity reduction, 38–40
Vulcanization, 43–44

W

Water, 1–3
  control gate, 4–5
  resources, 11–12
  retention, 28–29
  shortage, 11–12
Water conservation technologies, 11–12
  aerators for improving water quality, 20–22
  conservation of groundwater resources, 12
  conservation of surface water resources, 12
  ecological manifestation, 22–23
  fluid adopted for rubber dam, 20
  impact
    on environment and biodiversity, 23
    on society, 23
  installations of rubber dam for water conservation, 16–19
  rainwater harvesting, 13
  river morphology, 14
  rubber dam
    storage, 15–16

for water conservation and management, 13–14
Water-filled dam
operational steps for, 89–90, 90*f*
rubber dam, 5–9, 7*f*, 20, 49, 50*f*
Watershed programs, 12
Wave force, 63
Weight of rubber membrane, 55
Wetland perimeter of pond, 15

X
Xiaobudong rubber dam, 16

Y
Yihe river, 16, 17*f*, 18*t*

Z
Zoobenthos, 22
Zooplankton, 22
Zunkur SHP project, 113*t*
discharge values, 107*t*
initial installation cost, 107, 108*t*
LCC, 117*t*
NPW, 115*t*
revenue loss, 110*t*, 111*t*